Architectural Contract Document Production

ARCHITECTURAL CONTRACT DOCUMENT PRODUCTION

Thomas Berg
With Darla G. Berg

McGraw-Hill, Inc.

New York   St. Louis   San Francisco   Auckland   Bogotá
Caracas   Lisbon   London   Madrid   Mexico   Milan
Montreal   New Delhi   Paris   San Juan   São Paulo
Singapore   Sydney   Tokyo   Toronto

Library of Congress Cataloging-in-Publication Data

Berg, Thomas
    Architectural contract document production / Thomas Berg.
        p.   cm.
    ISBN 0-07-004857-6
    1. Architectural contracts—Documentation.   2. Communication in
architectural design   I. Title.
    NA2584.B47 1991
    720'.28'4—dc20   91-25130
                                                                        CIP

1 2 3 4 5 6 7 8 9 0   HAL/HAL   97654321

ISBN   0-07-004857-6

The sponsoring editor for this book was Joel Stein, and the production
supervisor was Donald Schmidt. It was set in Friz Quadrata and Times
Roman by Archetype, Inc.

Printed and bound by Halliday Lithograph.

Contents

Acknowledgments

We would like to acknowledge the many people and companies without whose help this book would not have been possible. We are especially grateful to the following:

Bassetti, Norton, Metler, Rekevics Architects in Seattle, Washington for their patience and generous cooperation during the production of this book. Olympic Blueprint Company in Seattle, Washington for their advice, their cooperation, and the sharing of their expertise which greatly aided in the production of this book.

The following supplied artwork, information, and other aides for the book:

American Standard Plumbing Company; Staedtler, Inc. in Chatsworth, California; Zipatone, Inc. in Hillside, Illinois; Chartpak, Inc.; Kroy, Inc. in Scottsdale, Arizona; Design Mates in Downers Grove, Illinois; Mayline Hamilton in Sheboygan, Wisconsin; Hewlett Packard Company in Palo Alto, California; Diazit in Youngsville, North Carolina; Tom, Donna, Shannon, and Sara Starbuck of Denver, Colorado; Heather Rash of Kennewick, Washington.

We would also like to acknowledge the patience and expertise given by the book's editor, Joel Stein, and by McGraw-Hill, Inc.

We would be remiss if we did not mention the incredible patience of our wonderful daughter, Mackenzie, without which this book would never have been written or typed, and for whom the effort was made in the first place.

For our joy, Mackenzie

List of Figures

GRAPHIC STANDARDS

1

INTRODUCTION

Graphic standards have been published by numerous authors and organizations, providing more than enough options which, unfortunately, confuse everyone. Add to this the fact that we, as architects, seem to enjoy designing graphic standards almost as much as we enjoy designing buildings, and you have a real problem.

Many attempts have been made to standardize production graphics, including those by the Northern California Chapter of the American Institute of Architects in their *Recommended Standards,* by the American National Standards Institute (ANSI) in numerous publications, and by Ramsey Sleeper in their *Architectural Graphic Standards.* Although these efforts were generally good, they each appeal to separate and distinct groups, and are therefore not universally accepted. The design and construction industry, as a result, has no true standard way to illustrate drawings.

One reason existing graphic standards don't work is a lack of needed symbols. Most of the libraries that I have seen aren't complete. For a graphic-symbol library to work, it must be as complete as possible. Brevity for the sake of saving space on a title sheet is the most frequently used reason encountered for reducing the symbol library. This results in continual redesign of graphic standards for each new project, and consequently, the loss of large amounts of time. Problems incurred by this brevity more than outweigh any benefits.

This chapter deals with graphic standards, abbreviations, and notations. I will define each graphic element and tell where and why it is used. I will also include some tips for reproducing the graphic standard on an economical title sheet that works for any project.

THE GRAPHICS LIBRARY

The symbols used in architectural drafting can be grouped under four main headings: *line type, lettering style, material legend,* and *graphic symbols.* Collectively, these make up the symbol library. Following are descriptions of each group with examples and illustrations.

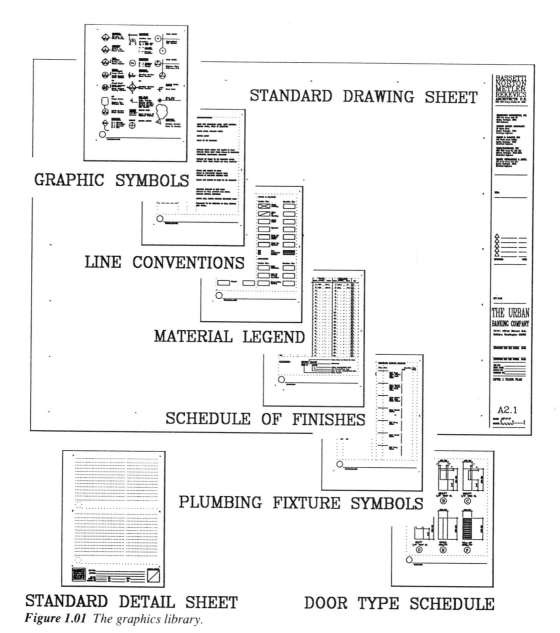

GRAPHIC SYMBOLS

STANDARD DRAWING SHEET

LINE CONVENTIONS

MATERIAL LEGEND

SCHEDULE OF FINISHES

PLUMBING FIXTURE SYMBOLS

STANDARD DETAIL SHEET

DOOR TYPE SCHEDULE

Figure 1.01 *The graphics library.*

LINES

Line work comprises the main portion of any drawing, and must be produced with care and skill to convey its meaning. Figure 1.02 illustrates a drawing with no attention to line quality. There is no distinction in width between object lines and dimension lines. There is no difference in configuration between visible lines and hidden lines. This drawing is hard to read. By adding the simple qualities of width and configuration to line work as in Fig. 1.03, the same drawing becomes much more clear.

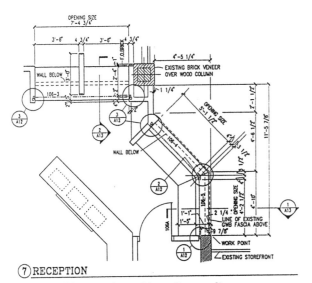

Figure 1.02 *Drawing without line quality.*

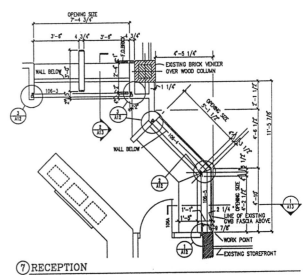

Figure 1.03 *Drawing with line quality.*

Line Width and Configuration

Line width and configuration can be varied in many ways to produce desired results. Figure 1.04 is a chart of recommended standards which can be read easily and accurately. Notice the specific design of each line type (configuration) and the mechanical pen point (line weight) to use for drafting.

LINE CONVENTIONS

LINE	PEN SIZE	PEN NO.	APPLICAICATIONS
────────────	0.25	3X0	USED FOR DIMENSION LINES, NOTE LEADERS, CEILING GRIDS, ITEMS IN ELEVATION
── ─ ── ─ ──			FLOOR LINES, COLUMN GRIDS
─ ─ ─ ─ ─ ─			HIDDEN LINES
- - - - - - -			ITEMS TO BE REMOVED
────────────	0.35	0	OUTLINE ITEMS ABOVE THE FLOOR IN PLAN OUTLINE ITEMS AWAY FROM WALLS IN ELEVATION FURNITURE, EQUIPMENT, FIXTURES
- - - - - - -			OUTLINE OF ITEMS TO BE REMOVED WHICH STAND AWAY FROM FLOOR OR WALL SURFACE LIKE CASEWORK, EQUIPMENT AND FIXTURES
────────────	0.50	1	WALLS AND DOORS IN PLAN WALLS IN REFLECTED CEILING PLAN OUTLINE IN BUILDING SECTION VIEW
▬ ▬ ▬ ▬ ▬			WALLS AND DOORS IN PLAN TO BE REMOVED
────────────	0.70	3	BUILDING OUTLINE IN SITE PLAN OUTLINE IN WALL SECTION AND DETAIL CLOUDS AROUND REVISIONS
▬▬ ▬ ▬▬ ▬ ▬▬			MATCH LINE, CLOUD AROUND ENLARGED PLAN
▬▬▬▬▬▬▬▬			ELEMENTS TO BE REMOVED IN WALL SECTION AND DETAIL

Figure 1.04 Line conventions.

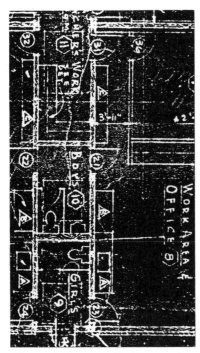

Figure 1.05 *Microfilm image of an old drawing.*

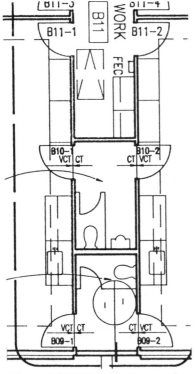

Figure 1.06 *Microfilm image of a CAD drawing.*

LETTERING

Lettering styles have changed a great deal over the past 50 or so years. Prior to the age of photographic reproduction and micro-imaging, architectural lettering was usually made to fit the space available on the drawing. This often resulted in letters less than $\frac{1}{16}$-inch tall. If you have ever tried to read a microfilm reproduction of one of those early drawings, you know how difficult it can be, and you can see the importance of good lettering at appropriate size.

Good lettering means readable lettering at reduced scale by poor reproduction processes. Much of our work is reduced to one-half its original size. At this reduced scale, every note and dimension should still be readable. To achieve this, lettering must be about $\frac{1}{8}$-inch tall. This minimum size is now, in fact, required by many government agencies.

This minimum-size letter is then used for dimensions, notes, and to fill in schedules and forms. Larger letters, say $\frac{5}{32}$-inch tall, are used for such things as room titles to make them stand out on a crowded plan drawing and for schedule column headings. Drawing and schedule titles should be lettered using $\frac{3}{16}$-inch-tall letters. This size also works well for sheet titles. If there is a need to use letters taller than $\frac{1}{4}$-inch, they should be produced mechanically. It is very difficult for most people to make a uniform letter taller than $\frac{3}{16}$-inch, and do it consistently. The result is frequently a shaky configuration of letters that looks a lot like a first grader's attempts to master the alphabet.

Hand Lettering

The following is a table of hand lettering. All lettering is vertical with a chisel point for accent of horizontal strokes. Sloped lettering of any kind should be avoided because it is difficult to get more than two people to do it the same way. The vertical style can be easily produced by using a small triangle as a guide for vertical strokes. This way, every person working on a drawing will produce similar and legible lettering. This lettering can also be reproduced by most CAD equipment, making hand lettering on CAD drawings indiscernible from that which is computer-generated.

HAND LETTERING GUIDE

2 X 4 WOOD STUDS @ 16' OC

1/8 inch tall lettering is used for the basic minimum lettering height
when required by local authorities. It is used for notes and dimensions,
and to identify furnishings and equipment.

AUDITIORUM
CORRIDOR

5/32 inch tall lettering is used for such things as room titles so that they
stand out in a crowded drawing. Use a soft pencil to produce dense lettering.

LEVEL 1 FLOOR PLAN
ROOM FINISH SCHEDULE

3/16 inch tall lettering is used for drawing titles, and for headers on schedules
such as the Room Finish Schedule. Use a strong stroke with a very soft pencil.

URBAN BANKING COMPANY
FIRST FLOOR PLAN

FIRST FLOOR PLAN

1/4 inch tall lettering should be reserved for special conditions like presentations.
If good quality hand lettering can not be achieved at this size, use mechanically
produced type.

Figure 1.07 *Hand-lettering guide.*

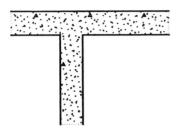

Figure 1.08 Concrete poche in plan view.

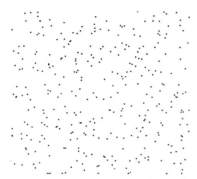

Figure 1.09 Concrete poche in elevation view.

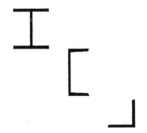

Figure 1.10 Steel at small scale.

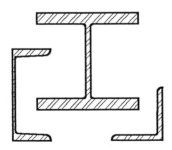

Figure 1.11 Steel at large scale.

MATERIAL LEGEND

Architectural drawings show two main views of a building. These are *section* and *elevation*. The way to show material in each of these views will not always be the same. For example, the graphic representation for a concrete wall in plan view (this is really a section looking down) is not the same as that for a concrete walkway (elevation view) on the same plan drawing. The *material legend* must take into account the need to show materials in both of these views. Also, for some materials a different symbol is used on large-scale and small-scale drawings.

Some material symbols have been around for a long time, and the construction industry is accustomed to seeing them represented on our drawings in the same old way. A good example of this is the symbol for concrete. For these old familiar symbols, it is best to stick with them as closely as possible; however, *Computer Aided Drafting* (CAD) must be considered when developing a material legend. Some old-time standards like the dot pattern are extremely labor-intensive for CAD and may take a great deal of time and computer memory to draw. When dots are used, they should be spaced as far as practical and still convey their meaning. The material legend which follows takes these conditions into consideration for over 40 commonly used materials. The legend is divided into section and elevation views, and is further arranged into related groups such as "Earth and Pavement," and "Concrete and Masonry."

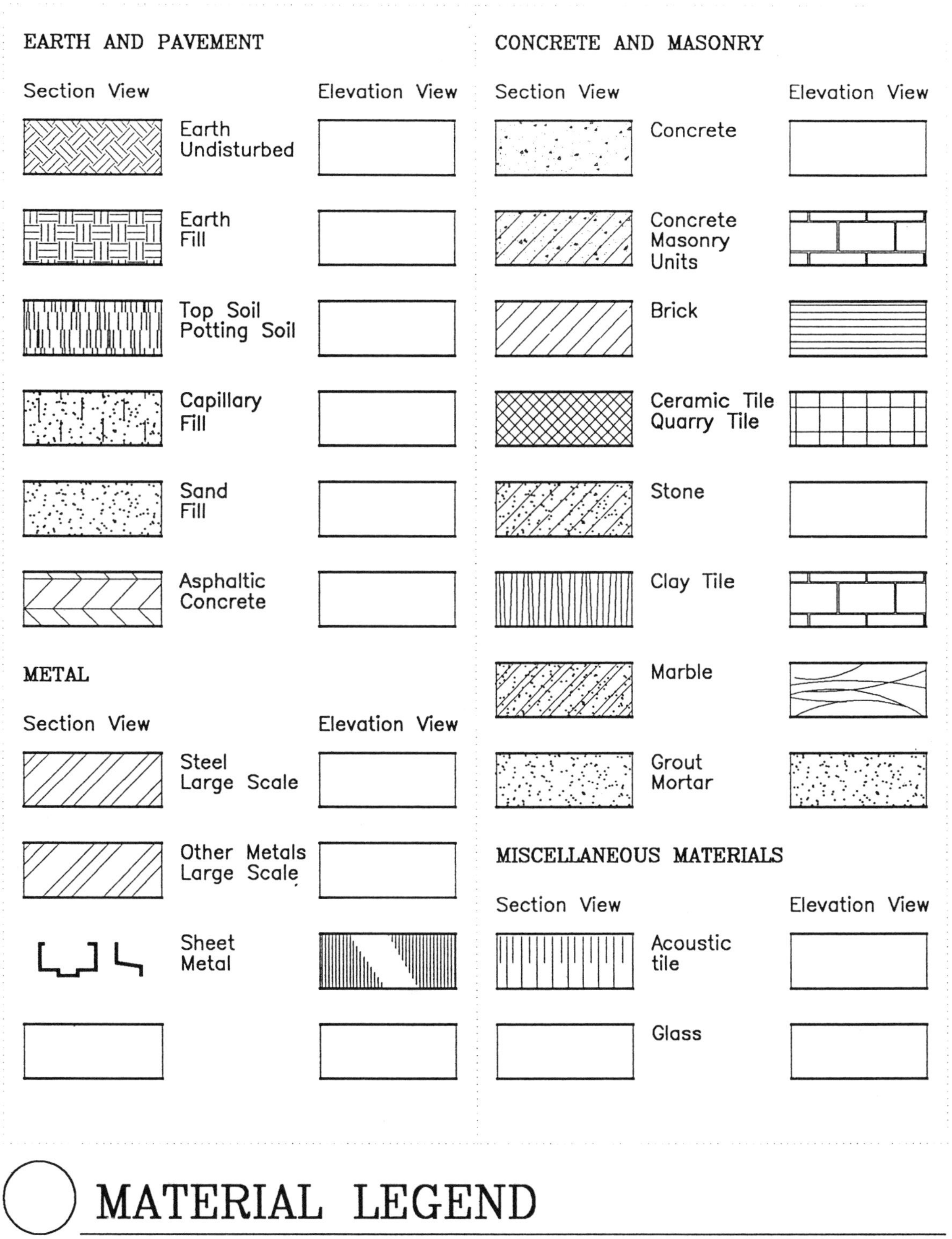

EARTH AND PAVEMENT

Section View · Elevation View

	Earth Undisturbed	
	Earth Fill	
	Top Soil Potting Soil	
	Capillary Fill	
	Sand Fill	
	Asphaltic Concrete	

METAL

Section View · Elevation View

	Steel Large Scale	
	Other Metals Large Scale	
	Sheet Metal	

CONCRETE AND MASONRY

Section View · Elevation View

	Concrete	
	Concrete Masonry Units	
	Brick	
	Ceramic Tile Quarry Tile	
	Stone	
	Clay Tile	
	Marble	
	Grout Mortar	

MISCELLANEOUS MATERIALS

Section View · Elevation View

	Acoustic tile	
	Glass	

MATERIAL LEGEND

Figure 1.12 Material legend.

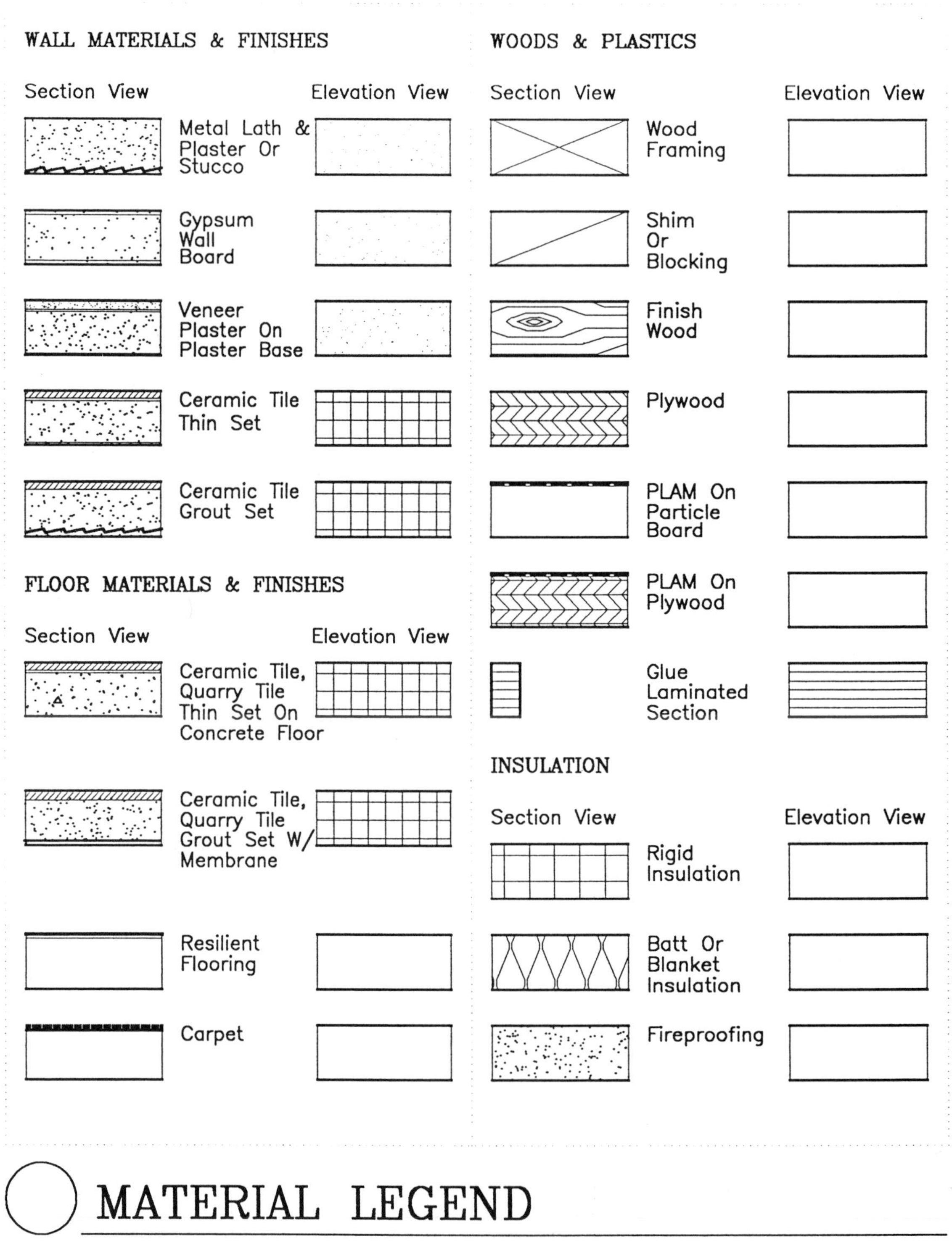

WALL MATERIALS & FINISHES

Section View | Elevation View
Metal Lath & Plaster Or Stucco

Gypsum Wall Board

Veneer Plaster On Plaster Base

Ceramic Tile Thin Set

Ceramic Tile Grout Set

FLOOR MATERIALS & FINISHES

Section View | Elevation View

Ceramic Tile, Quarry Tile Thin Set On Concrete Floor

Ceramic Tile, Quarry Tile Grout Set W/ Membrane

Resilient Flooring

Carpet

WOODS & PLASTICS

Section View | Elevation View

Wood Framing

Shim Or Blocking

Finish Wood

Plywood

PLAM On Particle Board

PLAM On Plywood

Glue Laminated Section

INSULATION

Section View | Elevation View

Rigid Insulation

Batt Or Blanket Insulation

Fireproofing

MATERIAL LEGEND

Figure 1.13 Material legend.

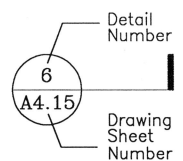

Figure 1.14 *Detail reference indicator.*

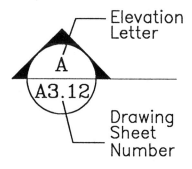

Figure 1.15 *Building elevation reference indicator.*

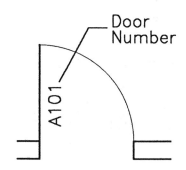

Figure 1.16 *Door symbol.*

GRAPHIC SYMBOLS

Graphic symbols have two basic functions in architectural drawings. These are to make reference to some other drawing or schedule, and to illustrate objects such as a chair, a door, or a sink. These two functions can be combined in the same symbol as in a door swing drawn in plan view. The graphic tells the reader it is a door, and the number tells the reader what kind of door it is and possibly much more (see schedules).

Symbols which refer a reader to another place in the drawings usually say two specific things: the drawing's unique number on a sheet, and the page or drawing-sheet number on which the drawing can be found. Notice the unique I.D. number is always on top, and the sheet number is always on the bottom. Do not mix up this order from one reference symbol to another, or the reader will begin to confuse the two.

Drawings are generally identified by a number in the reference symbol, but there are two exceptions. Both the *building elevation* and the *building section* are traditionally identified by a letter. The drawing-sheet number is still located on the bottom.

A variety of symbols have been invented to show the functions of reference and illustration. Some examples are doors, windows, partitions, plumbing fixtures, electrical devices, ceiling plan symbols, and symbols used in interior elevations. With these symbols, however, the reference number has a different meaning than it does with symbols like the section indicator. For example, the door-swing symbol contains only one number. This number is unique, and represents the door's location in the building by repeating the room number. This number, however, does not say where to find the information about the door. See Chap. 9, "Finding Your Way Through the Drawings," for further discussion about symbols.

Templates

Following are groupings of graphic symbols which illustrate all of the features discussed so far. The symbols were prepared by using CAD. The same results can be achieved by using inexpensive templates like the ones drawn below. Special shapes and proprietary templates should be avoided because they are not standard, and it is likely that someone in the office will not have them. The result would be a mixture of symbols on the same drawing, all trying to represent the same object. These two templates should be furnished to all staff architects within an office, and their use enforced.

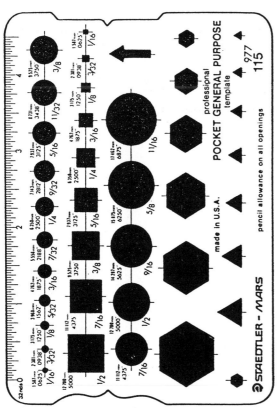

Figure 1.17 *Standard templates.* (Standard templates courtesy of Staedtler, Inc.)

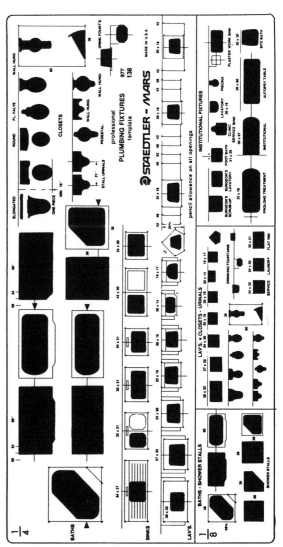

Figure 1.18 *Plumbing template.* (Plumbing template courtesy of Staedtler, Inc.)

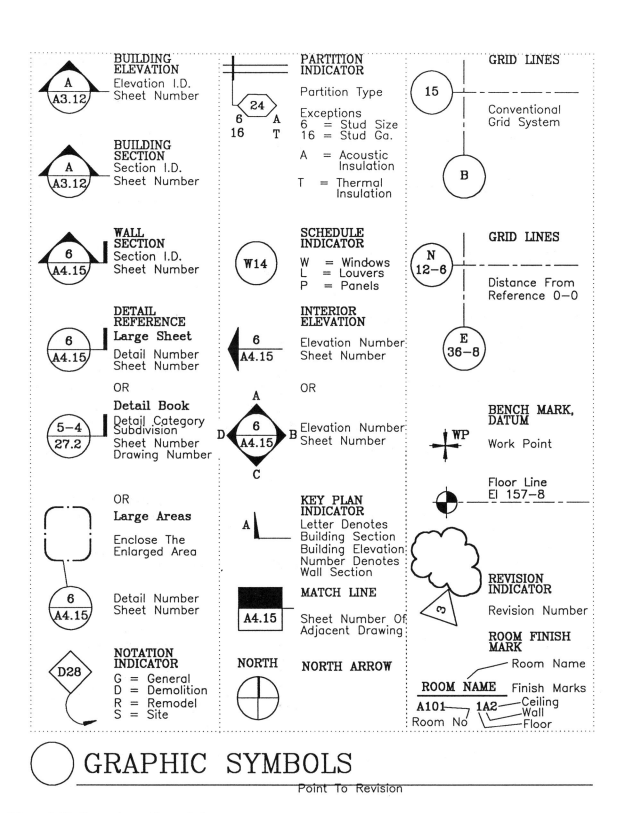

GRAPHIC SYMBOLS

Figure 1.19 General graphic symbols.

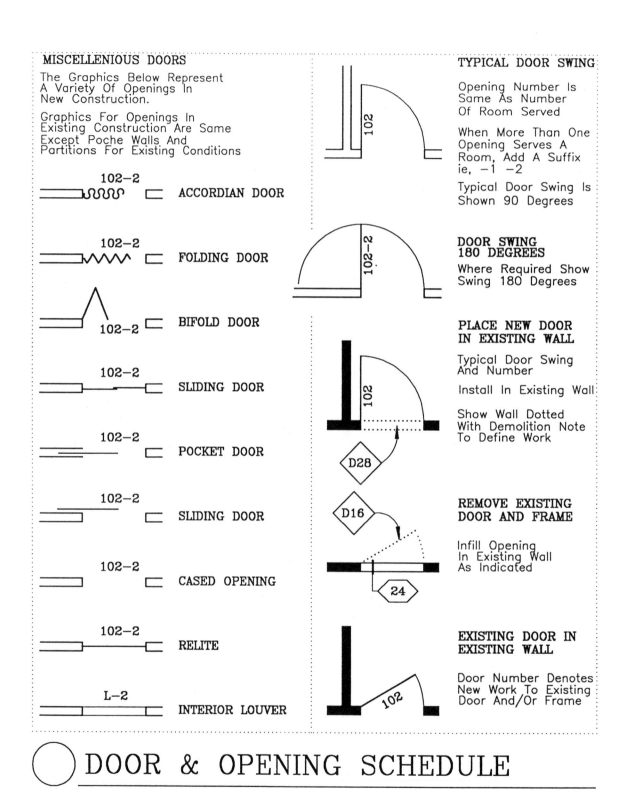

MISCELLENIOUS DOORS

The Graphics Below Represent
A Variety Of Openings In
New Construction.

Graphics For Openings In
Existing Construction Are Same
Except Poche Walls And
Partitions For Existing Conditions

102-2 ACCORDIAN DOOR

102-2 FOLDING DOOR

BIFOLD DOOR 102-2

102-2 SLIDING DOOR

102-2 POCKET DOOR

102-2 SLIDING DOOR

102-2 CASED OPENING

102-2 RELITE

L-2 INTERIOR LOUVER

TYPICAL DOOR SWING

Opening Number Is
Same As Number
Of Room Served

When More Than One
Opening Serves A
Room, Add A Suffix
ie, -1 -2

Typical Door Swing Is
Shown 90 Degrees

**DOOR SWING
180 DEGREES**

Where Required Show
Swing 180 Degrees

**PLACE NEW DOOR
IN EXISTING WALL**

Typical Door Swing
And Number

Install In Existing Wall

Show Wall Dotted
With Demolition Note
To Define Work

**REMOVE EXISTING
DOOR AND FRAME**

Infill Opening
In Existing Wall
As Indicated

**EXISTING DOOR IN
EXISTING WALL**

Door Number Denotes
New Work To Existing
Door And/Or Frame

DOOR & OPENING SCHEDULE

Figure 1.20 Door and opening symbols.

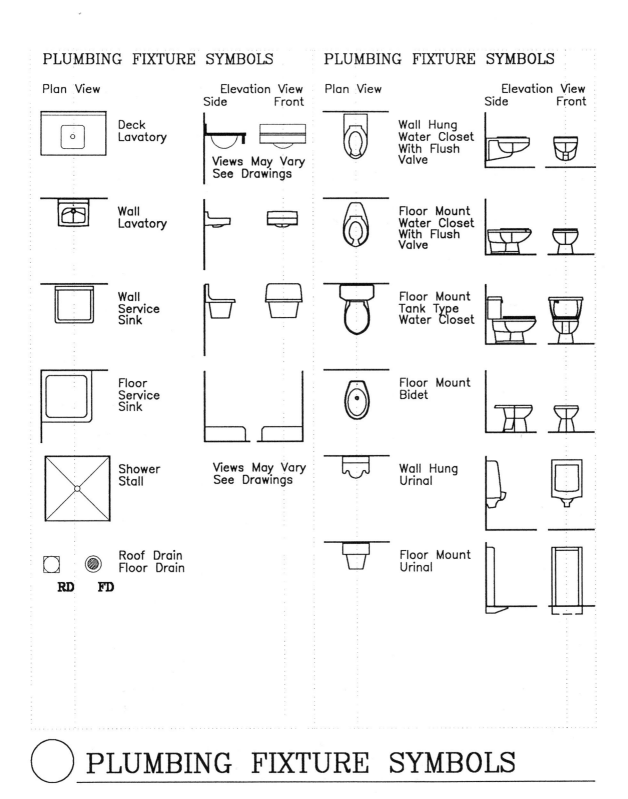

Figure 1.21 *Plumbing fixture symbols.* (Plumbing fixture symbols, some symbols courtesy of American Standard, Inc.)

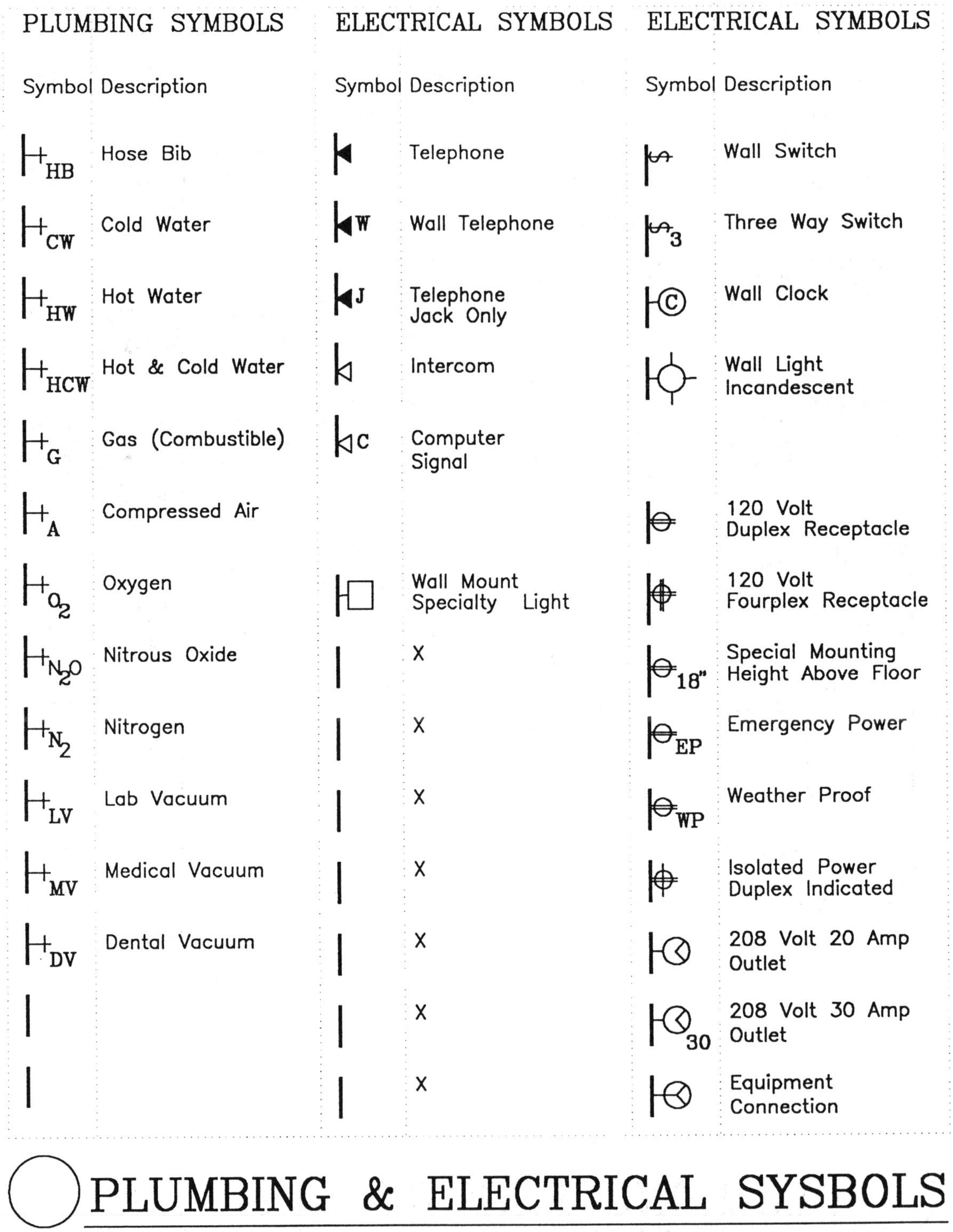

PLUMBING SYMBOLS		ELECTRICAL SYMBOLS		ELECTRICAL SYMBOLS	
Symbol	Description	Symbol	Description	Symbol	Description
$\vdash_{HB}$	Hose Bib	◀	Telephone		Wall Switch
$\vdash_{CW}$	Cold Water	◀W	Wall Telephone		Three Way Switch
$\vdash_{HW}$	Hot Water	◀J	Telephone Jack Only	⊢ⓒ	Wall Clock
$\vdash_{HCW}$	Hot & Cold Water	◁	Intercom		Wall Light Incandescent
$\vdash_{G}$	Gas (Combustible)	◁C	Computer Signal		
$\vdash_{A}$	Compressed Air				120 Volt Duplex Receptacle
$\vdash_{O_2}$	Oxygen		Wall Mount Specialty Light		120 Volt Fourplex Receptacle
$\vdash_{N_2O}$	Nitrous Oxide		X	$_{18"}$	Special Mounting Height Above Floor
$\vdash_{N_2}$	Nitrogen		X	$_{EP}$	Emergency Power
$\vdash_{LV}$	Lab Vacuum		X	$_{WP}$	Weather Proof
$\vdash_{MV}$	Medical Vacuum		X		Isolated Power Duplex Indicated
$\vdash_{DV}$	Dental Vacuum		X		208 Volt 20 Amp Outlet
			X	$_{30}$	208 Volt 30 Amp Outlet
			X		Equipment Connection

◯ PLUMBING & ELECTRICAL SYSBOLS

Figure 1.22 *Plumbing service and electrical symbols.*

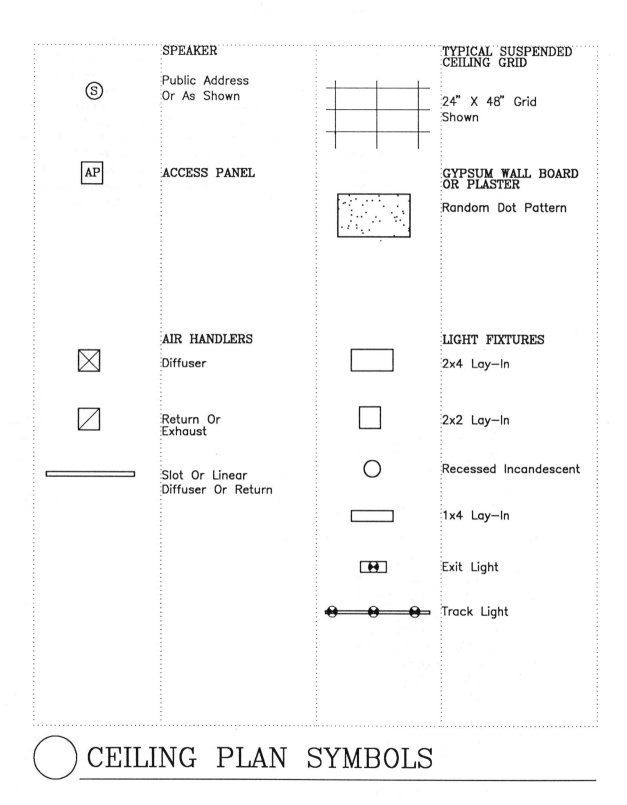

CEILING PLAN SYMBOLS

Figure 1.23 Ceiling plan symbols.

Publish Your Standards

Symbols for call outs are commonly shown on a cover sheet of graphic symbols. Often, however, the symbols used to illustrate objects are not included. Many calls from the contractor could be avoided by providing an accurate and complete list.

These symbols can be reproduced in a bound project manual or on large drawing sheets. For project manuals, simply insert copies of the individual graphic-standards pages into the book before binding. For large-sheet reproduction, arrange the artwork on a carrier sheet and have a negative shot at your local reprographics house. Keep this negative for blow-back on future projects. It can then be combined with other artwork to form a title sheet.

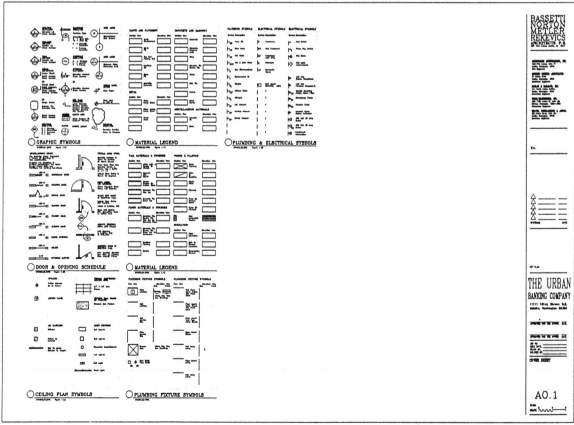

Figure 1.24 Graphic symbols on a cover sheet.

ABBREVIATIONS AND TERMS

In every major legal contract you will find definitions for the abbreviations and terms used within the contract. Lawyers like to leave as little room as possible for misinterpretation of the document. In architecture, the drawings and specifications are legal documents. As such they should be prepared with the same care as a lawyer uses in writing a contract. The abbreviations and terms used on architectural drawings should be defined at the front of each set of drawings just like an attorney does in a contract.

ABBREVIATIONS LIST

A

AB	Anchor Bolt
ACC	According/ Accordion
ACOUS	Acoustical
ACT	Acoustical Ceiling Tile
AD	Area Drain
ADD	Addendum
ADJ	Adjustable/ Adjacent
AFF	Above Finished Floor
AGG	Aggregate
AHU	Air Handling Unit
AL	Aluminum
ALT	Alternate/ Alteration
ANOD	Anodized
AP	Access Panel/ Apron Panel
APPROX	Approximate
APT	Apartment
ARCH	Architect/ Architectural
ASPH	Asphalt
AVE	Avenue
AVG	Average
AWP	Acoustic Wall Panel

B

BC	Bottom of Curb
BD	Board

Figure 1.25 *Partial abbreviations list.*

ABBREVIATIONS

Abbreviations are used on architectural drawings for two main reasons: to save space and to reduce the time it takes to do lettering. These reasons in themselves are fine, but the end product must always perform one more important function: to accurately and concisely convey a message to the reader. Here, as in any legal document, there is no room for ambiguity. The drawings will be read by a variety of people, including the owner, contractor, suppliers, and code officials, all with a varying ability to read and understand them. The abbreviations used must then be easily recognizable and match construction-industry standards, for example; "CONC" for concrete, and "GFRC" for glass-fiber reinforced concrete, and "EIFS" for exterior-insulation finish system.

Other things to consider when compiling an abbreviations list include the following:

Generally, words of four letters or less do not need an abbreviation. Some exceptions are "EA" for each, and "@" for at.

Keep the abbreviations list as short as possible. Do not include seldom-used abbreviations like "EST" for estimate, or abbreviations that are specific to a building type like "EXAM" for an examination room in a medical clinic. These should be placed in a supplementary abbreviations list at the end of the main list (unless you do a lot of medical work).

Stay with industry standards like "FHWS" for flat-head wood screw. Don't make up a new abbreviation without checking to see if one already exists.

PARTIAL ABBREVIATIONS LIST

INSP	Inspection
INST	Installation
INSUL	Insulation
INT	Interior

J

JAN	Janitor
JST	Joist
JT	Joint

K

KD	Kiln Dried
KIP	1000 pounds
KIT	Kitchen
KO	Knock Out
KP	Kick Plate
KS	Knee Space

L

L	Left/ Length
LAB	Laboratory
LAM	Laminated
LAV	Lavatory
LB	Pound

Figure 1.26 Partial abbreviations list.

ABBREVIATED FINISHES
ACOUSTIC CEILING SYSTEMS

AC-1   Exposed Tee Grid, 2'x 4'
         Lay-in Acoustic Panels.
AC-2   Exposed Tee Grid, 2'x 4'
         Lay-in Acoustic Panels
         1-hour Fire Rated.
AC-3   Exposed Tee Grid, 2'x 4'
         Lay-in Acoustic Panels.
AC-4   Exposed Tee Grid, 2'x 4'
         Lay-in Acoustic Panels
         1-hour Fire Rated.
AC-5   Concealed Grid, 1'x 1'
         Acoustic Tile.
AC-6   Concealed Grid, 1'x 1'
         Acoustic Tile.
AC-7   Adhesive Mounted, 1'x 1'
         Acoustic Tile

Note: A letter suffix (AC-4a) may be used to identify specific board and panel types.

Figure 1.27 *Partial abbreviations list.*

Sometimes a new abbreviation is necessary to fit the space provided. One such space limitation is on the *Room Finish Schedule* where columns are narrow. There may not be room to spell out "quarry tile," so "QT" is a good abbreviation.

Abbreviations can be used to represent entire finish systems such as "ACT-1," which can be defined as suspended 24″ × 48″ lay-in acoustical ceiling panels.

Do not include abbreviations that will appear in the specifications, such as agencies ("EPA"), testing laboratories ("UL"), and codes ("UBC"). A separate list of these abbreviations should be included in the specifications.

Do not use periods in abbreviations. They are unnecessary and a waste of time.

And last, if there is any doubt about using an abbreviation, forget it and write out the word or term being considered. Remember, the idea here is to communicate as accurately as possible, not merely to save space or time.

At the end of this chapter is a good abbreviations list to start with. It can be edited to suit specific office needs. Once it has been edited to suit your office and established as the standard, any additions should be added to the "Supplementary Abbreviations" list at the end.

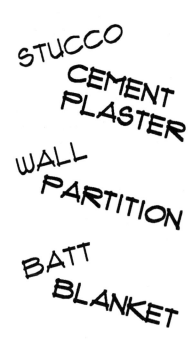

Figure 1.28 *Use the proper term.*

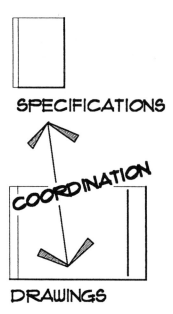

Figure 1.29 *Coordinate between drawings and specifications.*

TERMINOLOGY

A very important aspect of architectural drawing is *notation.* The words and terms used must be chosen with care and skill. For example, in dry-wall work it would be incorrect to call for "casing beads," as these devices are used in plastering. Another commonly misused term is "blanket insulation" which comes in 8-feet- and 10-feet-wide rolls for insulation between roof trusses and wall girts. Compare this with "batt insulation" which comes in sizes to fit standard 16-inch and 24-inch stud and joist spacing. One last set of words which are often misused is "wall" and "partition." A "wall" is a vertical plane enclosing a building, or that serves as an occupancy separation, and is generally load-bearing. A "partition," on the other hand, is a vertical plane in the interior of a building which subdivides space, and is generally not load-bearing.

It is important to get the right term so the reader can tell what is really meant. It is also important to coordinate the words and terms used by all persons working on the project. This includes the Specifications Writer. If the drawings show "stucco" and the specifications call for "cement plaster," you may get more or less than you anticipated. Stucco is used on the exterior, and cement plaster is used in the interior. Make sure the specification writer knows which application is needed.

At the end of this chapter is a list of words and terms which are often used in the preparation of architectural drawings. Recommended words and terms are shown boldface and capitalized. The comments in the right column are for assistance in understanding the word or term more clearly, and give some incorrectly used examples.

KINDS OF
NOTES

GENERAL
NOTES

SPECIFIC
NOTES

Figure 1.30 *Two basic kinds of notes.*

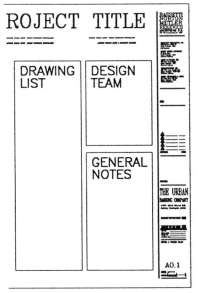

Figure 1.31 *General notes on the cover sheet.*

NOTATIONS

Notes used in architectural production bridge drawing and specifications by:

  ###   Telling the reader what a specific item is

  ###   Giving some kind of instruction

Notes can be placed to address a specific object on a drawing or in lists to give general information. This section deals with the use of notes to help the person reading the drawings to find the meaning intended by the architect. This chapter also defines the two types of notes. Refer to sections on abbreviations and terminology for acceptable items.

General Notes

There are two types of notes: *general notes* and *specific notes.* General notes are written either to preface the project as a whole, or to preface a section of drawings like window details. The list of general notes which addresses the whole project should appear on the title sheet where it is quickly found. General notes give specific instruction to the contractor that are very important and may not be shown elsewhere. The following is an example of a general note which helps the contractor find the way through the drawings:

> These drawings are arranged in chapters as shown on the Index to Drawings. See each chapter for additional notes which help explain the contents of the chapter.

This next example gives instructions about dimensioning:

> Partitions are dimensioned to the face of wallboard unless noted otherwise.

Notice the phrase "unless noted otherwise." This alerts the contractor to exceptions to the conditions expressed in the note.

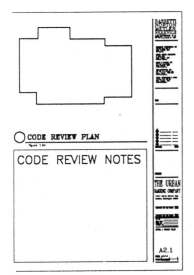

CODE REVIEW NOTES

Figure 1.32 *Code notes on the code plan sheet.*

General notes are not a place for saying things that belong in the specifications, i.e., "Paint all exposed metal." The specifications cover these kinds of items (or should), and do a much better job. Further, the general notes are not a place to describe conditions which should have been drawn before time ran out during contract document production. Usually, if the general notes are lengthy, the drawings are incomplete.

General notes are best used to preface individual sections of the drawings. If the project is divided into chapters (see Chap. 7), then each chapter should be prefaced by a series of notes which are aimed at helping the reader find his or her way. For example, the door section might be prefaced by this note:

   ###   Details called out in the Head, Jamb, and Sill columns of the Opening Schedule are found on sheet A-27.

In this manner, the schedule references only drawing 10 or 15 or 32, but does not need to repeat the drawing-sheet number each time.

Other notes are descriptive. The site plan is a location for such notes. There are the *legal description* and *zoning code notes* which explain how the project meets the requirements of the code. For example:

   ###   Allowable floor area is 24,000 gsf; actual floor area is 18,523 gsf.

This tells everyone that the building footprint is less than the maximum allowable.

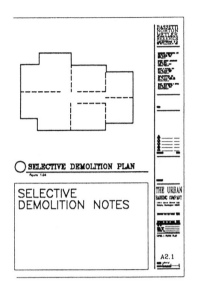

Figure 1.33 Selective demolition notes.

Demolition and Remodel Notes

For projects with demolition of existing structures or partial demolition of interior spaces, there are two sets of notes: *major demolition notes* and *selective demolition notes*. Major demolition notes address the massive demolition project. Selective demolition notes address more detailed problems. An example of a major demolition note is:

Remove existing concrete footings and foundations to Elevation 92.0'.

In an interior demolition job where existing space is being gutted to make way for a new work, the following note might be used:

Remove existing flooring to substrate leaving surface ready for installation of new finish scheduled.

Consider a project where selective demolition takes small bites out of an existing space leaving adjacent construction to be patched. A selective demolition note might read:

Remove existing metal stud and gypsum wallboard partition to limits shown on drawings.

When new construction patches into a demolition scar, the following note can describe the condition:

Patch remaining finishes in areas of new work to match existing.

A note like this can apply to gypsum wallboard, ceramic tile, or vinyl wall covering. One note works for all.

This note is used where an existing door or window is removed. You don't know the exact wall thickness, but want flush finish on each side of the patch. You also want no trace of the old door or window opening once the job is done.

Infill wall opening to match existing wall thickness and finish. Paint to nearest wall and ceiling breaks.

Plan General Notes

Floor plan notes can help the contractor find additional information like these examples:

Detail plans for stairs and elevators are found on sheet A-37.

Finish schedule is found on sheet A-26.

Other floor plan notes give construction directions like:

Partitions are dimensioned to face of gypsum wallboard unless noted otherwise.

The previous note sets the dimensioning point for the typical interior partition so it does not need to be stated each time a dimension is shown.

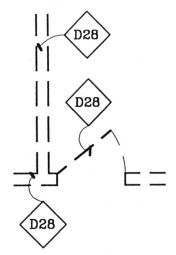

Figure 1.34 Notation symbols.

Graphic Presentation

Again, general notes are grouped into two kinds: those which help contractors find their way through the drawings and those which give specific instructions. These notes can be lengthy and are usually arranged into a list and numbered. *General notes* are numbered G01, G02, G03, etc. *Demolition notes* are numbered D01, D02, D03, etc. Where both major and selective demolition notes are listed, use D01 through D29 for major demolition, and divide selective demolition into floor and base conditions D30 through D49, wall conditions D50 through D69, and ceiling conditions D70 through D89. Leave the remaining D90 through D99 unused.

Each of these notes is used in the diamond symbol (see graphic standards, Chap. 1.) and placed on the drawings where they occur. *Remodeling notes* can be grouped similarly to demolition notes. *General remodel notes* should be numbered R01 through R29. Continue with *floor* and *base remodel notes* R30 through R49, *wall remodel notes* R50 through R69, and *ceiling remodel notes* R70 through R89. Again leave the last group, R90–R99 unused.

Specific Notes

The other basic kind of note is the *specific note*. This is the note that says something like: "2 × 4 stud at 16 oc." These notes are found on all drawings, and have the following properties:

They amplify the conditions found on the drawings.

They indicate material and size.

They clarify conditions not obvious in the detail such as "continuous" objects.

They note existing items. Everything is considered to be new unless specifically noted as existing.

They describe fasteners and anchoring conditions. (One of the more important notes placed on a drawing).

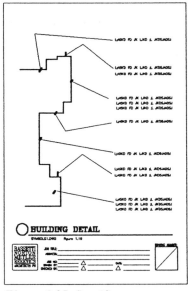

Figure 1.35 Specific notes.

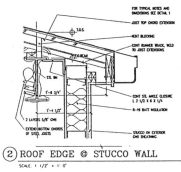

(1) TYPICAL ROOF EDGE @ CMU WALL
SCALE: 1 1/2" = 1' 0"

(2) ROOF EDGE @ STUCCO WALL
SCALE: 1 1/2" = 1' 0"

Figure 1.36 *Noting typical details.*

Typical Notes

Notes can be used to explain "typical" conditions once, so that it is not necessary to repeat them on every detail of similar situations. For example, in a complex roofing condition, place notes which apply to all roofing conditions on one detail and call them *typical.* On subsequent details write "see detail no._ for typical conditions." Then follow with notes specific to that detail.

NOTE: This practice should not be overused. For example, if a condition can be detailed in a few drawings, write out the repeat notes on all drawings. Use the typical note idea to save production time when a large number, say 15 or more details, are drawn and are really similar to the typical condition.

Here are some rules to observe when composing notes:

Always use standard abbreviations and terminology. Check terms with the specs.

Don't be wordy. Sometimes, the more you say, the deeper in trouble you get.

Do not repeat assemblies which are very clearly defined in the specifications. For example, in a ceiling suspension system simply note ACT-1, ACT-2, and let the specs explain the system.

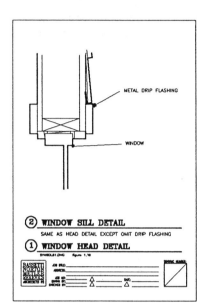

METAL DRIP FLASHING

WINDOW

② WINDOW SILL DETAIL
SAME AS HEAD DETAIL EXCEPT OMIT DRIP FLASHING
① WINDOW HEAD DETAIL
SYMBOLS1.DWG   figure 1.16

Figure 1.37 Window head detail.

The Word "Similar"

The word *similar* should be stricken from an architect's language. It has *no* meaning in contract document production. In its place use the term *same except* and explain the difference. For example, you are drawing a window-head condition and feel the sill is very much like it except there is no drip flashing. You don't want to spend the time to redraw the detail for the sill, so you say, "Sill, similar to head detail." If you don't explain the differences between the sill and the head conditions, the situation is open to interpretation. You may not get what you want. Instead of saying "similar," use the term "same except," and tell everyone, "Omit drip flashing and invert detail."

When noting a detail as being "same except opposite hand," be sure all the parts of the detail are manufactured opposite-hand. Many parts come in only one configuration and will not fit an opposite-hand detail.

CONCLUSION

To recap on notes, there are two basic types: *general* and *specific*. General notes can be grouped to preface important areas of drawings. Do not use extensive general notes at the beginning of the drawing set because this gives the appearance that the drawings are not complete and this could affect bidding. Be careful to use the same terms and abbreviations which are used in the specifications, and do not use the word *similar*.

STANDARD DRAWING SHEET

The largest single element of graphic standards is the *office drawing sheet*. It can be as small as 8½-inches-by-11-inches, or as large as 36-inches-by-48-inches. ANSI Y14.1 standards for engineering and government are shown below. Of these, the "A," "B," and "D" sizes are most often used. Other common sizes do not fit the ANSI standard, but are well suited to architectural production. These are:

11-inches-by-17-inches
30-inches-by-42-inches
34-inches-by-42-inches

ANSI Y14.1–1980
STANDARD
DRAWING SHEET
SIZES

POPULAR
DRAWING SHEET
SIZES

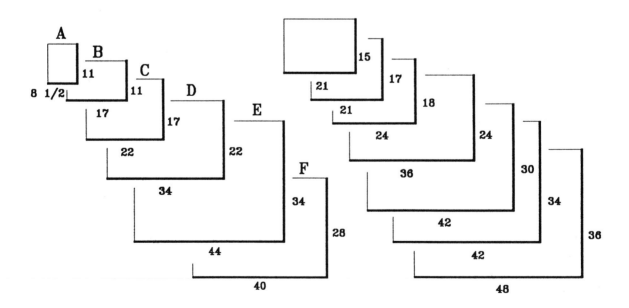

Figure 1.38 Standard drawing-sheet sizes.

Selecting a Drawing-Sheet Size

One time-honored way of selecting a drawing-sheet size is to see what size sheet can be used to fit the floor plan. If the plan fits on "D" size sheet with room for dimensions and title block, that's big enough. Some people feel that sheet size equates with project cost, i.e., the larger the sheet, the more expensive the building. In fact, this is not true. However, the psychological effect may be there, and this could result in higher bid prices. It is probably best to keep the sheet size as small as practical, even though you might need a couple more sheets to carry all the drawings.

With a variety of projects and floor-plan sizes, you could easily use every available sheet size. The next selection criterion is then based on storage space and economy. Storage should be considered for both paper and polyester film media, as well as reproduction stock such as blueline, blackline, paper sepia, and slick materials. The only places that have enough storage space for all these items are reprographics shops and vendors. So, the best combination of size to suit floor plan and economy will drive most practices to select at the most three sizes as "office standard."

Standard Drawing Module

8 1/2 INCHES

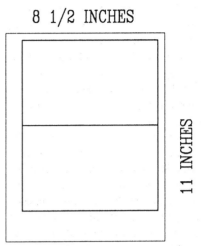

11 INCHES

Figure 1.39 Drawing-sheet composition.

There is one other criterion used to select drawing-sheet size. This is the manner in which information is placed on the sheet. Start with the "A"-size (8½-inches-by-11-inches) sheet. This is the most common size of drawing and writing medium available. It is, therefore, easy to store, and economical to use and reproduce. This sheet can be divided into two areas: one for title block and one for content. The space for content is set to the *standard drawing module* for all production work. Every drawing, note, and schedule should fit this module or a fraction or multiple of this module. By doing this, drawings can be saved, categorized, and filed for repeated use. (More on this in Chap. 6). The most efficient drawing module is 7¼-inches-wide-by-4½-inches-high with two modules fitting on an 8½-inches-by-11-inch sheet. This module is then arranged on the larger sheet.

42 INCHES

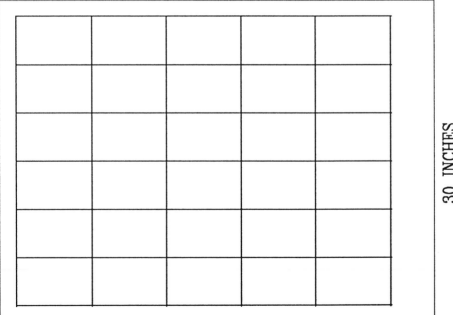

30 INCHES

Figure 1.40 The standard module.

The Title Block

Setting the standard module aside for a moment, the other area of concern on every drawing sheet is the *title block*. This area must contain enough information to precisely identify the sheet it is on from a stack of others. To accomplish this, the title block should contain these elements:

1 Name of the design firm doing the work:
This might be an individual, corporation, or joint venture. Include address, phone, and fax numbers. Make it big; it's free advertising.

2 Names of the major design consultants:
List civil, structural, mechanical, electrical, and any special consultants like acoustical, food service, etc. Include address, telephone, and fax number.

3 Space for professional stamps:
Some localities require the architect to stamp all drawings, even civil, mechanical, electrical, and structural drawings. If your local agency does not require the architectural stamp on engineering drawings, don't do it. That is what you pay the engineer for. Check with the agency to determine the need for stamping all drawings.

4 Revision log:
No one is perfect, so there must be spaces to enter the number and date of every revision made to a drawing sheet. Save room for at least six entries. Twice that many is not uncommon.

5 Key plan:
Provide space to draw a small key plan of the building. Use this plan to key in building elevation and section views relative to a sheet of drawings.

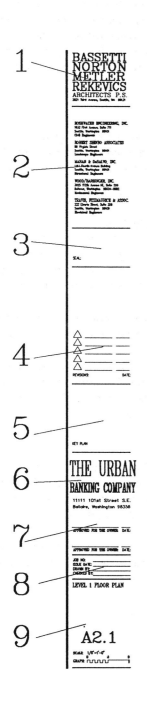

Figure 1.41 The title block.

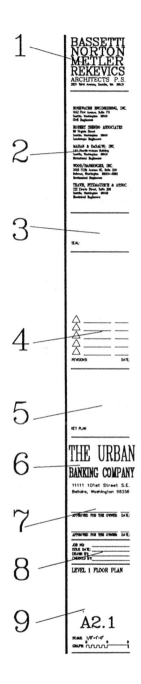

Figure 1.42 *The title block.*

6 Name of the project:

This is not the place to enter the owner's name unless specifically requested. Many owners don't want their names on the drawings. Conversely, some owners are very specific about the project title and how their name does fit into it. Always ask and obtain owner approval of the project name. Keep in mind that the project name must also include basic location information, i.e., street address, cross streets, etc.

7 Owner approval:

Provide a prominent location for the owner to sign approval of the document. Without this approval signature, a client can and often enough does require changes when it is too late to implement them without cost to the design team.

8 Project or job number, date, and credits:

Some projects become complex enough that they are divided into phases with an individual job number given to each phase. This individual job number is often one of few differences in the title block from one phase to another. Therefore, it is very important that it be entered accurately every time.

Every formal issuance of a drawing requires a date entry. Examples include phase submittals for owner approval and publication for bidding.

Provide space for the initials of the person doing the drawing and for the person checking the drawing. The intent here is not to point fingers but to allow others to know who did the work so questions can be directed to the right people.

9 Sheet title:

The drawing-sheet title is one of two handles by which a sheet is included into a set of drawings. Keep it short. Titles like "First Floor Plan" and "Finish Schedule" are fine. Do not list all the details drawn on the top edge of the sheet. (See Chap. 7 for additional information about drawing-sheet titles).

Below the title, enter the scale of the drawing on the sheet, and draw a bar graph of that scale. When more than one scale is used, draw a bar graph for each and enter "varies" on the scale line.

Another requirement of the title block is to make it easy to identify a roll of drawings without laying it flat. This generally means that the title block should be designed along the right-hand edge of the sheet. It can be placed on the bottom edge provided the job number, date, and sheet number occupy the right-hand end.

Drawing-Sheet Efficiency

Going back now to the basic sheet sizes, the standard drawing module is located to provide space for a title block along the right edge. This exercise will reveal the efficiency of each drawing-sheet size by showing how many modules a sheet can carry versus the amount of white space remaining, and allow a decision to be made on which size(s) to use as office standard.

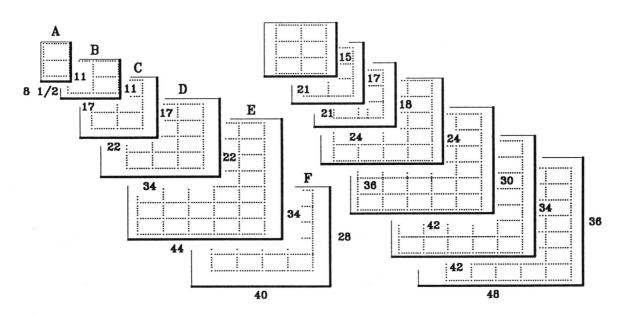

Figure 1.43 Evaluate drawing-sheet efficiency.

The 8¹/₂ × 11 Detail Sheet

There is one special case. The "A"-size (8½-inch-by-11-inch) sheet does not have enough space for a right-border title block. It is therefore best located only on the bottom edge. Space limitations may further restrict the amount of information this title block can contain. Design a title block that best displays the required information.

Figure 1.44 The "A"-size title block.

Design Your Standard Drawing Sheets

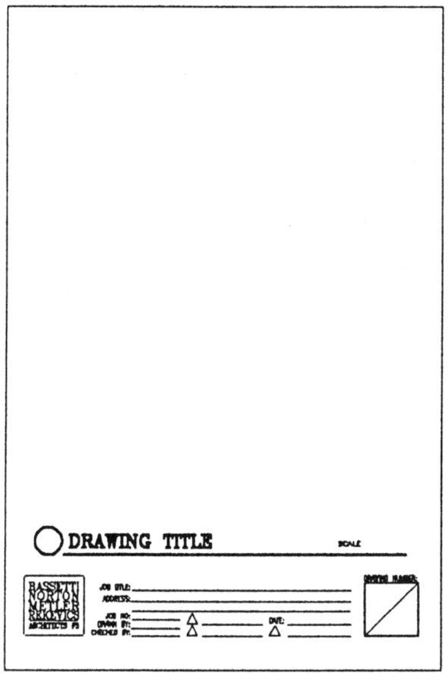

After reviewing these options, do the following:

Figure 1.45 *Standard 8^1/2-x-11 drawing sheet.*

Select the drawing-sheet sizes and configurations best suited to your practice.

Design a title block to suit the format that you have selected.

For manual production systems, keep preprinted drawing sheets on hand for all but the most preliminary work.

For CAD production systems, store the sheet formats for recall.

In offices where paper-storage space is at a premium, two sheet sizes might be enough. Consider 8½-inch-by-11-inch and 30-inches-by-42-inches with a vertical title block. These will yield the greatest economy overall for drawing production.

When more storage space is available, add larger sheets like 34-inches-by-42-inches.

The 30 × 42 Drawing Sheet

As you select a drawing-sheet format for an individual project, consider the number of details expected versus the number of larger drawings expected. As you have seen, some sheets are less efficient than others when drawing details, but they may be necessary to fit the floor plan and exterior elevations.

Last, the drawing sheet is a tool for producing drawings, and should accommodate any building configuration you can imagine. Always have enough options for drawing-sheet size and configuration to accommodate your designs.

Figure 1.46 *The standard 30 × 42 drawing sheet.*

RECAP

The graphic tools, symbols, abbreviations, and terms addressed in this chapter become the language used in architecture to define a building. That language must be consistent, at least within a set of drawings, and hopefully someday, within the industry.

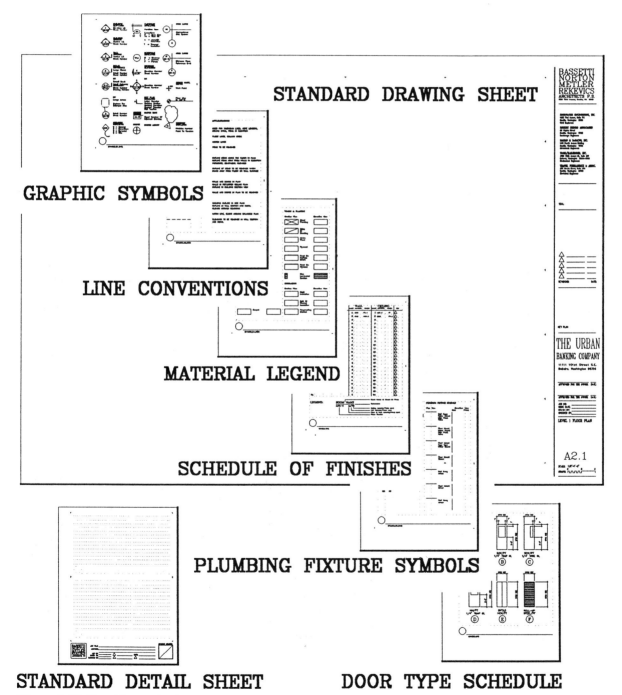

GRAPHIC SYMBOLS

STANDARD DRAWING SHEET

LINE CONVENTIONS

MATERIAL LEGEND

SCHEDULE OF FINISHES

PLUMBING FIXTURE SYMBOLS

STANDARD DETAIL SHEET

DOOR TYPE SCHEDULE

Figure 1.47 Graphic standards.

THE COVER SHEET

Graphic symbols and abbreviations can be produced on a cover sheet which is used for every project. Once the artwork is done, the sheet becomes a reusable tool for standardizing office production as well as an interpretative guide to aid the person reading the drawings.

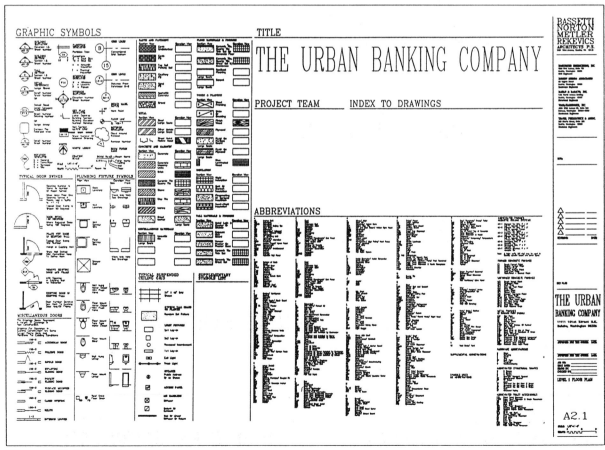

Figure 1.48 *Standard cover sheet.*

ABBREVIATIONS LIST

A

AB	Anchor Bolt
ACC	According/Accordion
ACOUS	Acoustical
ACT	Acoustical Ceiling Tile
AD	Area Drain
ADD	Addendum
ADJ	Adjustable/Adjacent
AFF	Above Finished Floor
AGG	Aggregate
AHU	Air-Handling Unit
AL	Aluminum
ALT	Alternate/Alteration
ANOD	Anodized
AP	Access Panel/Apron Panel
APPROX	Approximate
APT	Apartment
ARCH	Architect/Architectural
ASPH	Asphalt
AVE	Avenue
AVG	Average
AWP	Acoustic Wall Panel

B

BC	Bottom of Curb
BD	Board
BFD	Bi-Fold Door
BITUM	Bituminous
BL	Building Line
BLDG	Building
BLKG	Blocking
BLVD	Boulevard
BM	Beam/Bench Mark
BOT	Bottom
BRG	Bearing
BSMT	Basement
BTWN	Between
BUR	Built-Up Roof
BW	Bottom of Wall

C

C	Celsius/Centigrade
CAB	Cabinet
CAP	Capacity
CARP	Carpet
CB	Catch Basin/Chalkboard
CD	Ceiling Diffuser
CEM	Cement
CEM PL	Cement Plaster

CER	Ceramic
CG	Corner Guard
CH	Coat Hook
CHRL	Chair Rail
CI	Cast Iron
CIP	Cast In Place
CJ	Control Joint
CL	Center Line
CLG	Ceiling
CLKG	Caulking
CLO	Closet
CLR	Clear
CLSR	Closer
CMU	Concrete Masonry Units
CO	Clean Out
COL	Column
COMP	Composite/Composition
CONC	Concrete
COND	Condition
CONN	Connect/Connection
CONST	Construction
CONT	Continuous
COORD	Coordinate
CORR	Corridor
CR	Cold Rolled
CT	Ceramic Tile/Curtain Track
CTR	Center
CTSK	Countersunk
CU	Cubic
CW	Cold Water

D

d	Penny (nails)
db	Decibel
D	Clothes Dryer
DBL	Double
DECID	Deciduous
DEPT	Department
DET	Detail
DF	Drinking Fountain/Douglas Fir
DIA	Diameter
DIAG	Diagonal
DICA	Drilled In Concrete Anchor
DIFF	Diffuser
DIM	Dimension
DISP	Disposal
DL	Dead Load
DN	Down
DR	Door/Drain
DS	Downspout
DSP	Dry Stand Pipe
DW	Dishwasher
DWG	Drawing
DWR	Drawer

E

E	East
EA	Each
EB	Expansion Bolt
EJ	Expansion Joint
EL	Elevation
ELEC	Electrical
ELEV	Elevator
EMERG	Emergency
ENAM	Enamel
ENTR	Entrance
EP	Electrical Panelboard/End Panel
EQ	Equal/Earthquake
EQJ	Earthquake Joint
EQUIP	Equipment
ETR	Existing To Remain
EWC	Electric Water Cooler
EXC	Excavate/Excavation
EXH	Exhaust
EXP	Expansion
EXST(E)	Existing
EXT	Exterior

F

F	Fahrenheit
FA	Fire Alarm/Forced Air
FAB	Fabricate
FB	Flat Bar
FBD	Fiber Board
FD	Floor Drain
FDC	Fire Department Connection
FDN	Foundation
FE	Fire Extinguisher
FEC	Fire Extinguisher Cab
FF	Factory Finish
FG	Float Glass
FH	Flathead
FHC	Fire Hose Cabinet
FHMS	Flathead Machine Screw
FHWS	Flathead Wood Screw
FIC	Furnished and Installed by Contractor
FIN	Finish
FIO	Furnished and Installed by Owner
FIT	Furnished and Installed by Tenant
FL	Floor
FLASH	Flashing
FLUOR	Fluorescent
FM	From/Factory Mutual Research Corporation
FOB	Face Of Brick
FOC	Face Of Concrete
FOF	Face Of Finish
FOIC	Furnished by Owner Installed by Contractor
FTIC	Furnished by Tenant Installed by Contractor
FOIV	Furnished by Owner Installed by Vendor
FOM	Face of Masonry
FOS	Face of Studs
FP	Fireproof(ing)
FR	Freezer/Fire Retardant
FS	Full-Size/Floor Sink
FT	Foot or Feet
FTG	Footing
FURN	Furnace/Furnish
FURR	Furring
FUT	Future
FWP	Flat Wall Paint

G

G	Gas
GA	Gauge
GALV	Galvanized
GB	Grab Bar
GC	General Contractor
GD	Garbage Disposal
GFI	Ground-Fault Interrupter
GFRC	Glass Fiber Reinforced Concrete
GFRG	Glass Fiber Reinforced Gypsum
GFRP	Glass Fiber Reinforced Plaster
GL	Glass/Glazing/Glazed
GLAM	Glue-Laminated Wood
GND	Ground
GWB	Gypsum Wallboard
GYP	Gypsum

H

H	Hinge/High
HB	Hose Bib
HC	Handicap/Hollow Core
HCT	Hollow Clay Tile
HCW	Hot and Cold Water/Hollow-Core Wood
HDBD	Hardboard
HDNR	Hardener
HDR	Header
HDWD	Hardwood
HDWE	Hardware
HM	Hollow Metal
HORIZ	Horizontal
HP	Horsepower/High Point/Heat Pump
HR	Hour/Handrail
HS	Hook Strip
HT	Height
HTG	Heating
HW	Hot Water
HWR	Hot-Water Return

I

ID	Inside Diameter/Inside Dimension
IE	Invert Elevation
IG	Insulating Glass
IN	Inches
IND	Indicated
INFO	Information
INSP	Inspection
INST	Installation
INSUL	Insulation
INT	Interior

J

JAN	Janitor
JST	Joist
JT	Joint

K

KD	Kiln-Dried
KIP	1000 pounds
KIT	Kitchen
KO	Knock Out
KP	Kick Plate
KS	Knee Space

L

L	Left/Length
LAB	Laboratory
LAM	Laminated
LAV	Lavatory
LB	Pound
LH	Left Hand
LHR	Left-Hand Reverse
LKR	Locker
LL	Live Load
LP	Low Point
LSG	Laminated Safety Glass
LT	Light
LTWT	Lightweight

M

M/S	Mirror with Shelf
MACH	Machine
MAINT	Maintenance/Maintain
MAS	Masonry
MAT	Material
MAX	Maximum
MB	Machine Bolt/Marker Board

MC	Medicine Cabinet
MCW	Mineral Core Wood
MDF	Medium-Density Fiberboard
MDO	Medium-Density Overlay
MECH	Mechanical
MEMB	Membrane
MET	Metal
MEZZ	Mezzanine
MFR	Manufacturer/Manufacturing
MG	Mirror Glass
MH	Manhole
MIN	Minimum/Minute
MIR	Mirror
MISC	Miscellaneous
MO	Masonry Opening
MOD	Module/Modify
MS	Mirror with Shelf/Machine Screw
MTD	Mounted
MUL	Mullion

N

N	North
NO./#	Number
NOM	Nominal
NTS	Not-to-Scale

O

OA	Overall
OC	On-Center
OD	Outside Diameter
OFD	Overflow Drain
OFF	Office
OH	Oval Head
OHWS	Oval-Head Wood Screw
OPNG	Opening
OPP	Opposite
OSB	Oriented Strand Board
OZ	Ounce

P

P	Paint/Power
PA	Public Address
PB	Pegboard
PBB	Plaster Baseboard
PBD	Particleboard
PC	Precast
PCD	Paper Cup Dispenser
PERF	Perforated
PERP	Perpendicular
PG	Plate Glass
PH	Phase

PL	Plate
PLAM	Plastic Laminate
PLAS	Plaster
PLYW	Plywood
PNL	Panel
PNT	Paint
POL	Polish
PP	Push Plate
PR	Pair
PRE FAB	Prefabricate
PSF	Pound per Square Foot
PSI	Pounds per Square Inch
PT	Preservative Treated/Point/Post-Tensioned
PTD	Paper Towel Dispenser
PTDW	Paper Towel Dispenser & Waste Receptacle
PTN	Partition
PVC	Polyvinyl Chloride
PVMT	Pavement

Q

QT	Quarry Tile
QTR	Quarter

R

R	Riser
R&S	Backer Rod and Sealant
RA	Return Air
RAD	Radius
RB	Rubber-Base
RCP	Reflected Ceiling Plan
RD	Roof Drain
REBAR	Reinforcing Bar
RECEP	Reception
RECEPT	Receptacle
RECT	Rectangular
REF	Reference/Refrigerator
REG	Registration/Register
REINF	Reinforced
REQ	Required
RES	Resilient
RET	Retaining/Return
REV	Revision
RH	Robe Hook/Round Head/Right-Hand
RM	Room
RO	Rough Opening
ROW	Right Of Way
RP	Radius Point
RRL	Rub Rail
RS	Rough-Sawn
RWL	Rainwater Leader

S

S	South/Shelf
S&R	Shelf and Rod
S&V	Stain & Varnish
SAN	Sanitary
SC	Solid Core
SCD	Seat Cover Dispenser
SCHED	Schedule
SCR	Shower Curtain Rod
SCW	Solid Core Wood
SD	Soap Dish/Soap Dispenser
SEC	Second
SECT	Section
SF	Square Feet
SFC	Special Floor Coating
SGEN	Semigloss Enamel
SHR	Shower
SHT	Sheet
SHTG	Sheathing
SHV	Sheet Vinyl
SLR	Sealer
SMR	Sheet-Metal Raceway
SMS	Sheet-Metal Screws
SND	Sanitary Napkin Dispenser
SNW	Sanitary Napkin Waste Receptacle
SPEC	Specification
SQ	Square
SS	Service Sink
SST	Stainless Steel
ST	Stone
STA	Station
STC	Sound-Transmission Class
STD	Standard
STL	Steel
STOR	Storage
STRUCT	Structure/Structural
SUBFL	Subfloor
SURF	Surface
SUSP	Suspended
SWC	Special Wall Coating
SYM	Symmetrical

T

T	Top/Threshold/Tread/Toilet
T&B	Top & Bottom
T&G	Tongue & Groove
TB	Towel Bar/Tack Board
TBB	Tile Backer Board
TC	Terracotta
TCC	Top of Curb/Top of Concrete
TD	Towel Dispenser
TDW	Towel Dispenser & Waste
TEL	Telephone

TEMP	Tempered/Temporary/Temperature
TER	Terrazzo
TF	Top of Footing
TG	Tempered Glass
THK	Thick
THR	Threshold
TN	Toenail
TOD	Top of Deck
TOIL	Toilet
TOL	Tolerance
TOP	Top of Pavement
TOS	Top of Steel
TOW	Top of Wall
TPD	Toilet Paper Dispenser
TV	Television
TYP	Typical

U

UBC	Uniform Building Code
UC	Under Counter/Undercut
UL	Underwriters Laboratories, Inc.
UNFIN	Unfinished
UNO	Unless Noted Otherwise
UR	Urinal
UTIL	Utility
UV	Unit Ventilator/Ultraviolet

V

V	Vinyl
VAR	Variable/Varnish/Varies
VCT	Vinyl-Composition Tile
VERT	Vertical
VEST	Vestibule
VG	Vertical Grain
VOL	Volume
VP	Veneer Plaster
VT	Vinyl Tile
VWC	Vinyl Wall Covering

W

W	West/Water/Clothes Washer/WATT
W/O	Without
W/	With
WAIN	Wainscot
WC	Water Closet
WD	Wood
WDW	Window
WF	Wide Flange
WG	Wire Glass/Wire Gauge
WH	Water Heater
WIC	Woodwork Institute of California

WP	Work Point/Waterproof
WR	Waste Receptacle
WS	Weatherstripping
WT	Weight
WWF	Welded Wire Fabric
WWM	Welded Wire Mesh

Y

YD	Yard Drain/Yard

ABBREVIATED FINISHES
ACOUSTIC CEILING SYSTEMS

AC-1	Exposed Tee Grid, 2' × 4' Lay-in Acoustic Panels.
AC-2	Exposed Tee Grid, 2' × 4' Lay-in Acoustic Panels 1-hour Fire Rated.
AC-3	Exposed Tee Grid, 2' × 2' Lay-in Acoustic Panels.
AC-4	Exposed Tee Grid, 2' × 2' Lay-in Acoustic Panels 1-hour Fire Rated
AC-5	Concealed Grid, 1' × 1' Acoustic Tile.
AC-6	Concealed Grid, 1' × 1' Acoustic Tile.
AC-7	Adhesive Mounted, 1' × 1' Acoustic Tile

NOTE: A letter suffix (AC-4a) may be used to identify specific board and panel types.

FORMED CONCRETE FINISHES

F1	Rough-Formed Finish
F2	Smooth-Formed Finish
F3	Textured Finish
F4	Smooth-Rubbed Finish
F5	Sand Float Finish
F6	Sack/Grout—Cleaned Finish

UNFORMED CONCRETE FINISHES

U1	Rough-Screed Finish
U2	Scratched Finish
U3	Float Finish
U4	Nonslip Float Finish
U5	Drag Finish
U6	Broom Finish
U7	Ribbed Finish
U8	Exposed Aggregate Finish
U9	Dry-Shake Finish
U10	Salt-Seeded Finish
U11	Light-Steel Trowel Finish
U12	Full-Steel Trowel Finish
U13	Burnished Finish

HARDWARE ABBREVIATIONS

H	Hinge
G	Grouping
C	Closer
S	Stop
DP	Door Plate
T	Threshold
W	Weatherstripping

ABBREVIATED STRUCTURAL SHAPES

W	W Shapes
S	S Shapes
M	M Shapes
C	American Standard Channel
MC	Miscellaneous Channel
HP	HP Shape
L	Angle
WT	Structural Tee Cut from W Shape
ST	Structural Tee Cut from S Shape
MT	Structural Tee Cut from M Shape
PL	Plate
ST	Structural Tubing

ABBREVIATED TOILET ACCESSORIES

PTD	Paper Towel Dispenser
PTDW	Paper Towel Dispenser & Waste Receptacle
WR	Waste Receptacle
RH	Robe Hook
HS	Hook Strip
TPD	Toilet Paper Dispenser
SND	Sanitary Napkin Dispenser
SNW	Sanitary Napkin Waste Receptacle
MIR	Mirror
M/S	Mirror with Shelf
SCR	Shower Curtain Rod
SD	Soap Dish/Soap Dispenser
CH	Coat Hook
GB	Grab Bar
TB	Towel Bar
PCD	Paper Cup Dispenser
MPU	Multipurpose Unit
SCD	Toilet Seat Cover Dispenser

COMMONLY USED TERMS

The list below contains terms commonly used in architectural contract document production. Terms listed in boldface type are considered acceptable, and are followed by helpful amplifying information. Terms listed in normal-face type are not acceptable and should be avoided. Alternatives are given in the amplifying text.

ACCESS DOOR	Use only for small doors not included in the door or opening schedule.
ACCESS FLOOR	Not "pedestal floor" or "computer floor."
ACOUSTICAL PANEL	Lays into suspended ceiling grid.
ACOUSTICAL SEALANT	Sealant used in acoustical assemblies to isolate materials. See also **SEALANT.**
ACOUSTICAL TILE ADHESIVE	Adhesive applied to ceiling or other surface. Not "cement," "paste," "glue," or "mastic."
ALTERNATE	Formal description of work for purpose of obtaining cost.
ALTER, ALTERATION	Not "remodel."
anchor bolt	Use **EXPANSION BOLT.**
ANODIZE	Not "alumilite™," "alodize™," "kalcolor™," or "duranodic™," etc.
BACKER ROD	Round, rope-like joint filled. Used behind sealant.
BATT INSULATION	Roll-type insulation for installation between studs or joists. May be friction-fit or stapled-in.
bituminous	Not used.
BLANKET INSULATION	Roll-type insulation for installation over suspended ceilings and on smooth wall surfaces. May be laid loose between roof trusses or attached with stick clips.
BUILT-UP ROOFING	Not "tar and gravel" or "asphalt roofing."
bulkhead	Not used. See **RETAINING WALL.**
CAULK, CAULKING	Used only indoors in rare occasions where little movement is expected. Use **SEALANT.**
CASING BEAD	A plaster stop. Do not use in gypsum wallboard construction. See **METAL TRIM.**
CEMENT	Use for "Portland cement." Do not use for "glue" or "adhesive."

CEMENTITIOUS UNDERLAYMENT	Used to level out minor imperfections in floor prior to application of finish material. See also **GROUT, UNDERLAYMENT.**
CEMENT PLASTER	An interior Portland cement plaster. See also **STUCCO.**
CERAMIC TILE	Not "mosaic tile" or "wall tile."
CHALKBOARD	Not "blackboard."
COLD JOINT	Used where no bonding is expected or required between to surfaces of the same material, as in a concrete wall.
CONCRETE MASONRY UNITS	Not "concrete block" or "cinder block."
CONSTRUCTION JOINT	Used where a joint is required due to construction processes, but bonding of adjacent materials is also required.
CONTROL JOINT	A tooled cut or kerf used to control the location of future cracks.
DADO	A rectangular groove cut in the side of a board. See also **RABBET.**
DAMPPROOFING	A coating used to resist vapor transmission. Not intended to be "waterproofing."
DELETE	To remove from the contract documents. See also **OMIT** and **SUPERSEDE.**
DOWNSPOUT	A sheet metal or plastic conduit for rainwater. See **RAINWATER LEADER** when using pipe or tubing.
drywall	Not used. See **GYPSUM WALLBOARD.**
earth	Use **SOIL.** See also **TOPSOIL.**
ELASTOMERIC	A rubbery material used in connection with sealant, flashing, membrane, expansion-joint covers, etc.
EXPANSION JOINT	A joint designed for movement, both expansive and contractive. Do not use to describe a **CONSTRUCTION JOINT** or a **CONTROL JOINT.**
EXPANSION BOLT	Single-unit bolt with integral anchoring device, such as "Wej-it™" or "Kwik Bolt™." Do not use "anchor bolt" for these.
EXPANSION SHIELD	Use for devices that receive a separate screw or bolt and also note type of screw or bolt. Do not use "anchor bolt" for these. Not "cinch anchor."
felt	In built-up roofing, use **PLY.** For sheathing, use **BUILDING PAPER.**

FIBER GLASS	Not "fiberglass™."
FIREPROOFING PLASTER	Not "vermiculite™," "perlite™," etc.
flakeboard	Use **PARTICLEBOARD, ORIENTED STRAND BOARD.**
FLOOR SINK	Use for sink recessed flush in floor, or where indirect waste is required. Not a "service sink."
formica™	Proprietary, use **PLASTIC LAMINATE.**
free access floor	Use **ACCESS FLOOR.**
furnish and install	Don't use. Save this one for the specifications. Use **PROVIDE.**
furred ceiling	Not used.
FURRING	Not "stripping."
FURRING CHANNEL	Cold-rolled steel channel. For hat-shaped 25-gage steel channels see **METAL FURRING.** See also **RUNNER CHANNEL.**
GLASS STOP	Not "glazing bead."
GLAZED OPENING	Used at interior partitions. Sometimes called "borrow light," or "relite." See also **VISION PANEL.**
glue	See **ADHESIVE.**
GROOVE	A long, narrow indentation. In wood, use only when parallel to the grain. See also **RABBET.**
GROUT	Any cementitious material used to fill, level, or set other materials. Do not use "sulphur," "por-rock™," etc. to describe such materials. See also **CEMENTITIOUS UNDERLAYMENT.**
GUARDRAIL	A protective barrier where a floor drop-off exceeds code minimum. See also **RAILING.**
GYPSUM WALLBOARD	A variety of gypsum-based board products for interior and exterior uses. Not "drywall".
HANDRAIL	Use for a single rail against a wall. For protective barricade at open side of stair, use **RAILING.** See also **GUARDRAIL.**
HANGER	Any suspended structural member from which other members are attached.
HARDBOARD	A compressed and glued wood product available in sheets. Not "masonite™." See also **PARTICLEBOARD** and **ORIENTED STRAND BOARD.**

HARDWOOD	Note specific species. Wood from broad-leaved evergreen or deciduous trees. See also **WOOD** for softwood.
hat channel	Use **METAL FURRING** and detail "hat" shape.
HOISTWAY	Use for elevators and dumbwaiters. Not "shaft."
HOISTWAY BEAM	Beams supporting guiderails between multiple hoistways. Not "separator beams."
HOLLOW METAL	Use to describe doors and frames made from bent sheet metal, usually 18 to 16 gage. Do not use for extruded shapes. For pressed metal frames as in an access panel, see **METAL FRAME.**
INSTALL	Use instead of "apply," "hang," "place," etc. For "furnish and install," use **PROVIDE.**
INSULATING CONCRETE	A lightweight aggregate concrete used to provide both insulation and slope to roofs. Not "vermiculite™ concrete," "foam concrete," etc. Do not use for other **LIGHTWEIGHT AGGREGATE CONCRETE** which is not intended to provide insulation.
KICK PLATE	Armor plate on door face to take abuse. Not "toeboard."
kick space	Use **TOE SPACE.**
laminated plastic	Use **PLASTIC LAMINATE.**
LEADER	A rainwater conduit made of pipe or tubing. See **DOWNSPOUT** for sheet metal or plastic systems.
LIGHTGAGE METAL FRAMING	Weldable load-bearing metal studs and joists, usually 18 gage and heavier. For non-load-bearing framing of any gage for gypsum wallboard, use **METAL STUD** and **METAL FURRING.**
LIGHTWEIGHT AGGREGATE CONCRETE	Concrete of lightweight aggregate. Not concrete designed to provide insulation. See **INSULATING CONCRETE.**
masonite®	Is proprietary, use **HARDBOARD.**
mastic	Do not use. Use **ADHESIVE.**
METAL FURRING	Hat-shaped, and C-shaped steel channels of various gages used to furr out walls and ceilings. For cold-rolled steel channels see **FURRING CHANNEL.**
METAL LATH	Not "diamond mesh," "chicken wire," etc.

METAL STUD	Non-load-bearing interior metal framing assemblies. For weldable load-bearing assemblies use **LIGHTGAGE METAL FRAMING.** Not "screw stud," "C stud," etc.
METAL TRIM	Edge trim used for gypsum wallboard construction. Do not use "casing bead."
MINERAL FIBER	Not "asbestos."
moisture barrier	Use **DAMPPROOFING** or **VAPOR RETARDER.**
OMIT	To leave out by intention. See also **DELETE** and **SUPERSEDE.**
open-web joists	Use **STEEL JOISTS.**
OPTION	Use when contractor has a choice with no cost impact. See also **ALTERNATE.**
ORIENTED STRAND BOARD	Not "flake board." See also **PARTICLEBOARD.**
PANELING	Sheet or board material for interior use.
PANELS	Sheet material, with some sort of joint or trim for exterior or interior use. Any panelized building material such as precast concrete.
PARTICLEBOARD	A compressed and glued-wood fiber material produced in sheets. Not "chipboard." See also **HARDBOARD** and **ORIENTED STRAND BOARD.**
PARTITION	Non-load-bearing vertical panel subdividing interior spaces. Can have fire or acoustical ratings. For load-bearing conditions see **WALL.**
PATCH	Replacement or repair of material or finish to match existing conditions.
pedestal floor	Use **ACCESS FLOOR.**
PLASTER	Specifications define type, i.e., gypsum plaster, Keenes cement plaster. See also **CEMENT PLASTER** and **STUCCO.**
plaster stop	Use **CASING BEAD** for plaster. See also **METAL TRIM** for gypsum wallboard.
PLASTIC LAMINATE	Not "Formica™"; "laminated plastic."
plastic membrane	Use **DAMPROOFING.** Not "visqueen™." See also **VAPOR RETARDER.**
PLYWOOD	If several types of plywood are used within a drawing, coordinate with specifications and identify consistently.

PRECAST CONCRETE PANELS   Shop-cast or site-cast concrete elements which must be lifted and anchored into place. Not "tilt-up concrete," "exposed aggregate concrete," or "architectural precast."

PROTECTION BOARDS   Used to protect waterproofing from backfill material.

PROVIDE   Denotes "furnish and install," not "supply," "deliver," "erect."

QUARRY TILE   Not "paver tile."

RABBET   Groove on edge of a member only.

RAILING   Protective barricade at open side of stair. Not "balustrades." See also **HANDRAIL** and **GUARDRAIL.**

rainwater leader   Use **LEADER** if made of pipe or tubing. Use **DOWNSPOUT** if made of sheet metal or plastic.

RECORD DRAWING   Drawings revised by architect to include construction changes. Not "as-built drawing."

REFINISH   To put a finish back into its original condition. Do not use "refurbish," "rehabilitate," "remodel," "renew," or "renovate."

refurbish   Use **REFINISH.**

RELOCATE   To move from one location and install in another location.

remodel   Use **ALTER** or **ALTERATION.**

renovate   Use **REFINISH.**

REPLACE   To provide a substitute. Must be followed by name of replacement.

RESILIENT BASE   Not "rubber base," "vinyl base," etc.

RESILIENT FLOORING   See also **RESILIENT TILE, SHEET FLOORING, SEAMLESS FLOORING.**

RESILIENT TILE   Not "vinyl tile," "V.A.T.," "linoleum," etc.

restore   Use **REFINISH.**

RIGID INSULATION   A variety of insulation products that come in sheets. Not "fiberboard," "board insulation," "cellular insulation."

ROOF HATCH   For access to roof. Not "scuttle." See also **SMOKE VENT.**

RUNNER CHANNEL	The 1½-inch cold-rolled steel channel generally used as the main runner in suspended ceiling systems. See also **FURRING CHANNEL.**
SAME EXCEPT	Used to compare one object to another. Always followed by what is different. Never use "similar."
SCREED	Metal or wood strip used to gage thickness of applied materials such as plaster.
SCRIBE	Strip of material to make tight closure to adjoining surfaces as in a counter top to a wall. Not a "filler."
SEALANT	Use for joints subject to movement or weather penetration outdoors or indoors. If purpose is acoustical use **ACOUSTICAL SEALANT.** See also **CAULK.**
SEAMLESS FLOORING	Sheet material with joints field-welded or sealed, or field-installed materials finished to provide a homogenous flooring material. See also **SHEET FLOORING.**
SECTION	Drawing showing cut through an object. Not a rolled-steel shape.
SELF-EDGE	Plastic laminate applied to edge of plywood or particleboard using same material as surface material.
separator beams	Use **HOISTWAY BEAMS.**
SERVICE SINK	Wall or floor mounted sink. Not "janitor's sink" or "slop sink." See also **FLOOR SINK.**
SHEATHING	Any board or panel used to cover and protect the framing members of a wall. May also add strength or insulation to the wall system. Not "sheeting," "diaphragm," "plyscore."
SHEET	Thin construction material, not a drawing.
SHEET FLOORING	Resilient flooring installed in lengths, generally wall-to-wall with joints depending upon manufactured widths of roll material. See also **SEAMLESS FLOORING.**
SHEET METAL	Use general term on drawings. Specifications should define particulars.
sheet rock®	Proprietary, do not use. See **GYPSUM WALLBOARD.**
SMOKE VENT	Use specifically for venting smoke. See also **ROOF HATCH.**

similar	Do not use. See **SAME EXCEPT.**
SOIL	Not "earth," "dirt," etc. See also **TOPSOIL.**
sound caulking	Use **ACOUSTICAL SEALANT.**
SOUND DEADENING BOARD	High-density wallboard, wood fiber or gypsum, not suitable for painting or finishing.
SOUND RETARDANT	Not "soundproof."
STAGGER	To offset elements in a horizontal or vertical plane, as stagger studs, stagger bolts.
STEEL JOISTS	Not "open-web joists," "bar joists," or "truss joist."
STONE	Use for all stone and composition stone products. Not "granite," "slate," "marble," etc. Define type, quality, etc. in the specifications.
stripping	Use **FURRING.**
structural studs	Use **LIGHTGAGE FRAMING.**
STUCCO	An exterior, three-step, Portland cement plaster. See also **CEMENT PLASTER.** Do not use for synthetic finish systems. See also **SYNTHETIC STUCCO.**
STUD	Upright framing member of wood or metal. See **LIGHTGAGE METAL FRAMING** and **METAL STUD.**
SUBFLOORING	Usually of different grade and thickness than used for wall or roof sheathing. Not "sheathing" or "subfloor."
SUPERSEDE	Used to indicate that a drawing, detail, etc. is being replaced by another. Followed by the name of the new item. See also **OMIT** and **DELETE.**
SYNTHETIC STUCCO	May be used as a finish coat over cement plaster base, or in system with rigid insulation and other materials, "exterior insulation and finish system."
TACKBOARD	Not "bulletin board;" "corkboard."
THRU	Acceptable short version of *through* on drawings only.
tilt-up concrete	Don't use. This is a method of site casting and erection. Use **PRECAST CONCRETE.**
TOEBOARD	Raised protective edge at balconies, landings, etc. (OSHA requirement).
TOE SPACE	Recess at base of cabinets. Not "kick space."

TOPSOIL	Use only in landscape construction. See also **SOIL.**
TYPICAL	Representative example, characteristic of a kind.
UNDERLAYMENT	A smooth, hard material placed over rougher substrates to achieve a surface suitable for the application of such finishes as resilient flooring. See also **CEMENTITIOUS UNDERLAYMENT** and **PARTICLEBOARD.**
VAPOR RETARDER	A sheet or coating which impedes the flow of water vapor through a wall, floor or ceiling assembly. Not "visqueen," "plastic film," "waterproofing," or "house wrap."
VISION PANEL	A glazed opening in a door.
WAINSCOT	Wall finish which does not extend to the ceiling.
WALL	Vertical element enclosing a building or that serves as an occupancy separation. Generally load-bearing. See also **PARTITION.**
wallboard	Use **GYPSUM WALLBOARD.**
WATERPROOFING	Designed to resist a head of water. Not "membrane," "dampproofing," "waterproof," etc.
WOOD	Use only for solid stock softwoods. See also **HARDWOOD, PLYWOOD,** etc.

(The preceding list of terms is reproduced from the POP Manual and edited with permission of The Northern California Chapter American Institute of Architects.)

ANATOMY OF A DRAWING

2

INTRODUCTION

The main purpose of any architectural drawing is to convey a message. For example, a floor plan defines the shape of a building and the arrangement of interior spaces. A building elevation shows building massing and fenestration. Detail drawings show how individual parts and pieces come together for a specific purpose. The techniques used to prepare these drawings can be as varied as their contents and with as many interpretations.

Chapter 1 dealt with the graphic elements used in contract document preparation. This chapter deals with the way in which these graphic tools are used in solving a drawing problem.

Drawing Clarity

Architectural drawings are made up of graphic symbols, dimensions, and notes, each placed in a drawing in a specific location to convey a very deliberate message. It is when this process is not fully understood that poor drawings with little meaning are produced. The example in Fig. 2.01 shows a window 2'-2"-wide-by-3'-6"-high. It is impossible to tell from this drawing whether the dimensions refer to the size of the window or the rough opening. By simply adding the designation of "WD" for "Window Dimension" and another string of dimensions marked "RO" for "Rough Opening," the drawing becomes much more meaningful.

It is important to think about every element placed on a drawing and to judge its usefulness and accuracy. The following anatomy of a drawing deals with individual elements of a drawing and gives specific reasons for their use.

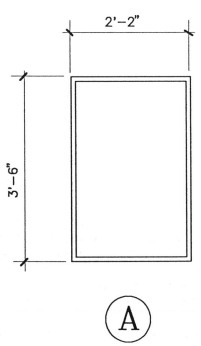

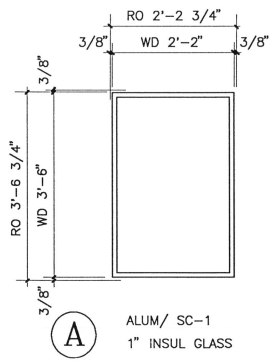

Figure 2.01 *Window type with insufficient information.*

Figure 2.02 *Same window type with sufficient dimensional information.*

DRAWING SCALE

The first decision to make with any drawing is to select the scale. The *Drawing Scale Selection Chart* lists drawing types and the smallest and largest scales normally used for each. For example, the window in Fig. 2.02 would be drawn at $\frac{1}{4}'' = 1'\text{-}0''$ scale.

DRAWING SCALE SELECTION CHART

DRAWING	MINIMUM		MAXIMUM	COMMENTS
Site Plan	1 : 30		1 : 20	Site Plan and Survey should be the same size.
Detail Site Plan	1/8"	to	1/2"	For large areas.
	1 1/2"	to	3"	For small area details.
Floor Plans, Exterior Elevations & Building Sections	1/16"			For very large buildings.
			1/8"	For general use.
Detail Plans	1/4"	to	1/2"	For large areas.
	1 1/2"	to	3"	For small area details.
Building Elevations	3/8"	to	3/4"	For detail elevations.Use same scale as selected for wall sections.Elevations
Wall Sections	3/8"			Keep small scale drawings sketchy.
			3/4"	Large scale sections might reduce the number of details, but don't count on it.
Interior Elevations	1/4"			Typical scale for 95% of this work.
	3/8"	to	3/4"	For lelvations with lots of detail and/or finish conditions. Use same scale as selected for wall sections.
Reflected Ceiling Plans	1/8"	to	1/4"	Use same scale as selected for floor plans.
Details	1 1/2"	to	3"	Select one scale or the other for all details in a family.
Schedules, Doors, Windows, Cabinets	1/4"			Typical scale.
	3/8"	to	3/4"	For complex items. Select one scale for all drawings in a family.

Figure 2.03 *Drawing scale selection chart.*

Floor Plan Scale

Floor plans of large structures, like an apartment building, might be drawn schematically at $\frac{1}{16}'' = 1'-0''$, and the typical units redrawn at $\frac{1}{4}'' = 1'-0''$. This process saves many drawing sheets and the time it would take to draw repetitive apartment plans at large scale.

In medical work, such as a neighborhood clinic, the entire building plan should be drawn at $\frac{1}{4}'' = 1'-0''$. This practice will provide space on the drawing for needed notes, detail references, and dimensions.

Another consideration when selecting drawing scale is the local building department. Some agencies require that all lettering be a minimum height of $\frac{1}{8}''$. With lettering this large, the drawings often need to be produced at a larger scale. For this reason, you may find that $\frac{1}{4}'' = 1'-0''$ plans are required for most building types.

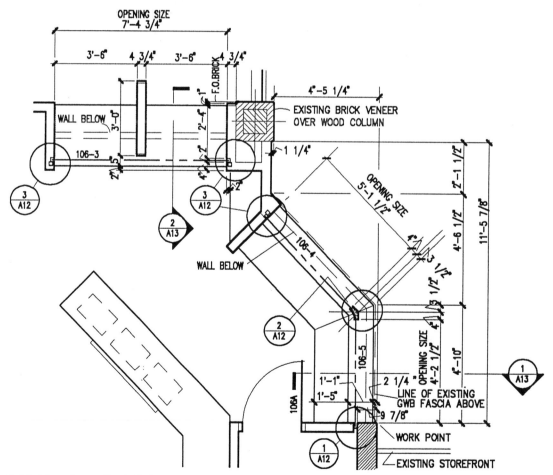

Figure 2.04 *Plan of a medical clinic.*

Maintain Relative Scale

It is important to keep the same scale for all drawings which are related. For example, if the floor plans of a building are drawn at ⅛″ = 1′-0″, then all building elevations should be drawn at ⅛″ = 1′-0″. To change scales to ¼″ = 1′-0″ would be confusing. Similarly, if the floor plans are drawn at a scale of ¼″ = 1′-0″, the building elevations should also be drawn at ¼″ = 1′-0″. This way, the person reading the drawings will "see" the whole building at the same relative scale, and not have to make mental interpolations when paging from plans to elevations.

The same idea holds for details. Select either 1½″ = 1′-0″ or 3″ = 1′-0″ for all exterior details. This goes for roofing details as well as plaza-paver details. This subtle rule will help the reader visualize all components of the building at the same relative size.

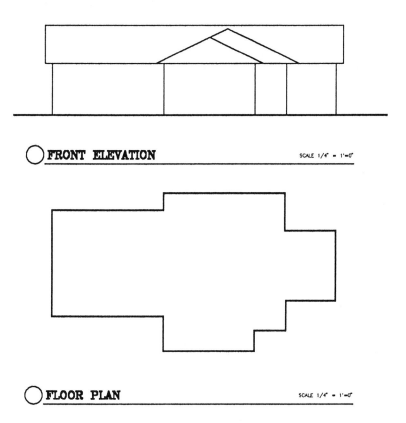

Figure 2.05 *Produce all building plans and elevations at the same scale.*

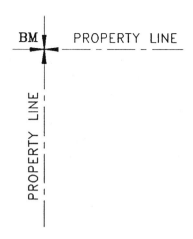

Figure 2.06 Bench mark.

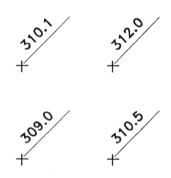

Figure 2.07 Spot elevations on a site plan.

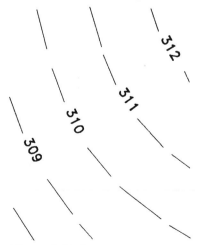

Figure 2.08 Contours on a site plan.

DIMENSIONING

Dimensioning is the language of size. It tells how big something is, as in a 2-×-4 wood stud, or measures distance from one object to another, like 12'-6". Dimensions can be used to instruct the contractor where to put an object, to compute building area, or to record the as-built condition.

Site Dimensions

Dimensioning for a project usually begins with the building site. A survey drawing is made by a civil engineer along with a legal description. These documents record the size, topography, vegetation, and any existing structures on the site. A reference point is established and called the *bench mark* or *datum*. This mark is the beginning point for every dimension that follows. From it, the contractor locates the building both horizontally and vertically.

Dimensions are typically given in feet and decimals of a foot. Vertical dimensions are shown as spot elevations at an interval of 10.0 feet on center (more or less depending on site slope) or by contours. Contours are much easier to read, but may have a 50 percent error from contour to contour.

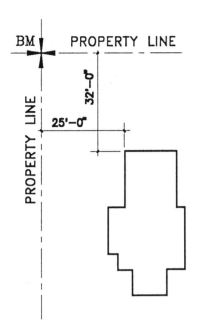

Figure 2.09 Locate the building on the site plan.

Locate the Building

The building should be dimensioned from one, and only one, point: the bench mark. In some cases this bench mark might not be on the site, so the next choice is the property corner nearest the bench mark. Run out a dimension from the bench mark along two perpendicular property lines to the points where the building is to be located. Here, place dimensions to the building which run perpendicular to the property lines. This will set the building on the site. If the building does not sit square with the site, or the site is not square, additional dimensions will be needed to locate the building. Use the same property lines for such additional dimensions.

The dimension line shown on the site plan represents the "face of the building." This line can be the actual face of brick on a flat wall surface, or some reference line, when the building surface has subtle variations.

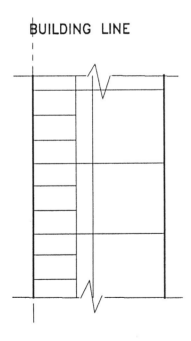

Figure 2.10 Building line can be actual face of building.

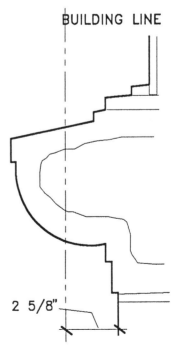

Figure 2.11 Building line can be a reference line for complex wall shapes.

Grid Lines

Within the building, major reference lines, called *grid lines,* placed perpendicular to each other, are drawn through all column centers. This structural grid is numbered in one direction and lettered in the other direction. This grid of lines, along with the face of building lines, become the horizontal-dimensional control for the building.

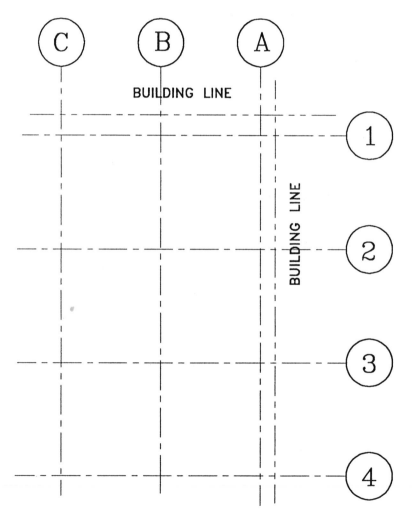

Figure 2.12 *Column grids and building lines make up horizontal dimensional control.*

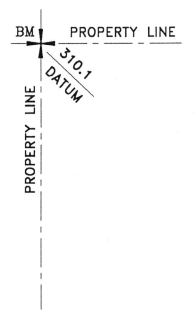

Figure 2.13 Datum.

Floor and Roof Lines

Vertical-dimensional control begins with the datum, out on the site. The first floor is often set as reference elevation 100'-0" which is further defined as being equal to a desired site elevation of, say, 256.75'. From this point on, all vertical distances are established. The main reference lines are the floor lines. They may represent the actual finished surface of the floor, but are more likely to represent a reference line that may also equal the top of the structural floor. Many finishes will extend above this level. It is helpful to prepare a drawing that shows the actual elevation of floor finishes relative to the reference floor line.

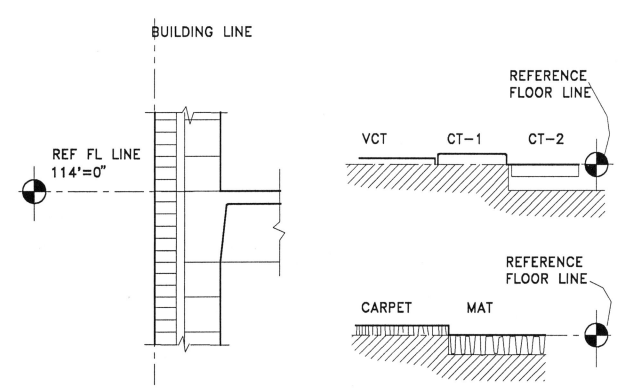

Figure 2.14 Reference floor lines.

Figure 2.15 Finishes relative to reference floor line.

Roof-Reference Line

The reference line used to locate a roof is not so easy to establish. In the case of a flat structure, the reference roof line will also be the top of structural roof. The finished roof will slope and is located by spot elevations at high points and low points, and along edges and walls. In the case of a sloping structure, two reference lines are needed to define the slope, one at high point and one at low point. Spot elevations are again used to define slopes.

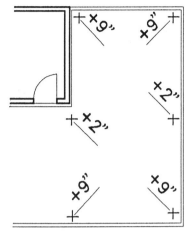

Figure 2.16 *Roof plan with spot elevations.*

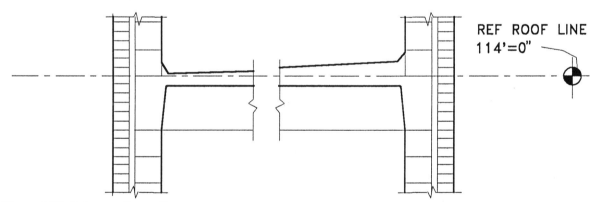

Figure 2.17 *Reference line at flat structures.*

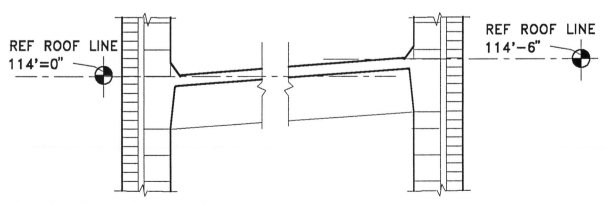

Figure 2.18 *Reference lines at sloped structures.*

The Building Grid

This grid of horizontal and vertical lines becomes the spine of the drawing set. All drawings, especially details, must have at least one of these reference lines as its foundation.

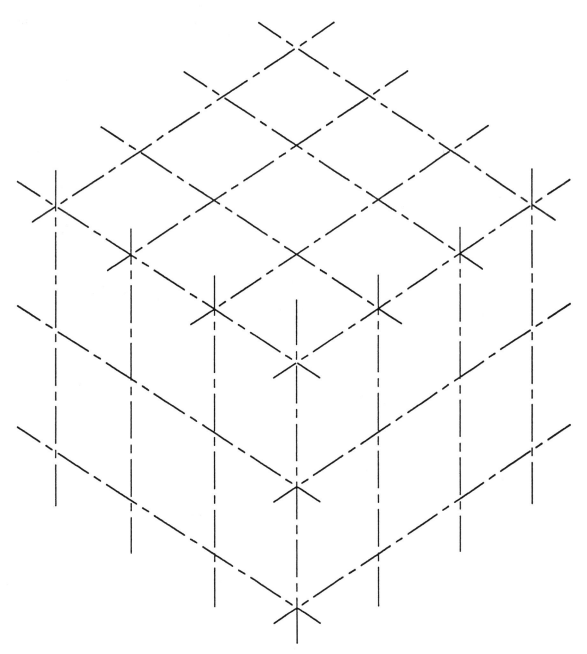

Figure 2.19 *Building reference lines.*

Plan Dimensioning Exterior

Floor-plan dimensioning starts on the exterior with the *overall building dimensions.* This is the total length of building along a front or side. Next, dimension wall breaks. This string will dimension building massing elements. Next, dimension the structural grid from the building line to grid and closing on the last building line. The last major dimension on the outside is for wall openings. Locate windows, doors, and storefronts from a grid or face of building line. When showing a dimension for doors or windows, be sure it matches the dimensions shown on the door or Window Schedule. This last dimension line can also be used to show dimension from grids to building wall breaks and other elements.

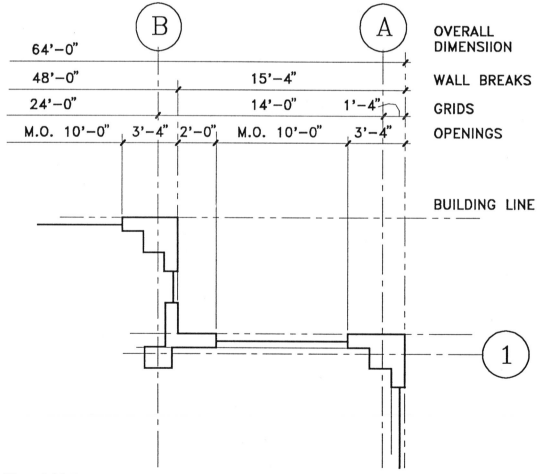

Figure 2.20 *Dimensioning the building exterior.*

Plan Dimensions Interior

Dimensioning on the interior begins with locating stud partitions. There are three basic schools of thought regarding partition dimensioning. These are:

1. Use partition centerline.
2. Use face of stud.
3. Use face of wallboard.

There are reasons for each, but it is not advisable to mix the systems on the same job.

Centerline

System number 1, *partition centerline,* is the system of dimensioning which originates with residential construction. Its main advantage is that it allows the layout contractor to locate plumbing rough-ins directly from plan dimensions without interpolating for wall thickness. It is also useful for placing partitions on window mullion centerline and column centerline. Its main downside is its inability to account for aligning a wall face when the partition thickness changes.

Face of Stud

System number 2, *face of stud,* is the system favored by most contractors who lay out partition framing. They can stretch a 100-foot tape and tick off the left face of all partitions that cross the dimension line. The resulting mark on the floor plate is the face of runner track or floor plate. It works fine when all partitions are the same thickness, but may cause alignment problems when a one-hour wall abuts a two-hour wall, or a shaft wall. If care is not taken, a room can end up ½″ too small because of wallboard thickness, resulting in insufficient space for the kitchen refrigerator or other important items.

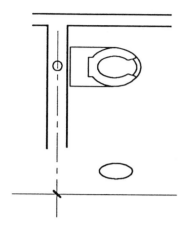

Figure 2.21 Centerline dimensioning is used to locate in-wall plumbing and electrical services.

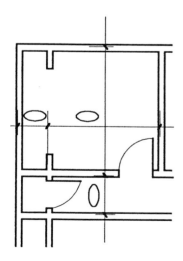

Figure 2.22 Face of stud dimensioning is easy to use and is preferred by wood framers.

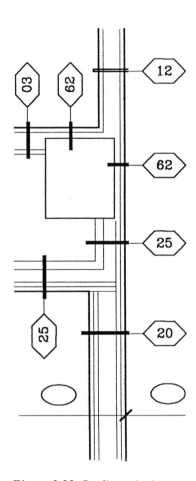

Figure 2.23 By dimensioning to face of wallboard, partition construction and stud location can vary without affecting the dimensions.

Face of Wall

System number 3, *face of wallboard,* is used for 95 percent of all work, and shows everyone what the finished product must be. A room dimensioned as 12'-0"-by-10'-0" will be exactly that. With system 1, it might be 11'-7½"-by-9'-7½". With system 2, it might be 11'-10⅜"-by-9'-10⅜", or some other size less than 12'-0"-by-10'-0". This could be very critical if stock cabinetwork is being ordered for a specific dimensioned room and the room ends up too small. The rule here is to always dimension to the face of wallboard. Then, if two layers of board are needed at a stair shaft and only one layer of board is needed at a corridor, one dimension can clearly describe both cases. In the case of a wall with ceramic tile over tile backer board, dimension to the face of tile backer board. The tile is finish material and may encroach into the room space. As in the condition of reference floor line, the face of wallboard becomes a "reference" line for adding finish. In this way, a dimension to a partition will allow wallboard to align without regard to finish. Ceramic tile, wood paneling, and paint are all given the same importance. The downside to this system is that the layout contractor must interpolate from the Partition Schedule to find face of stud for layout work. The upside to this, however, is that it places the responsibility for accurate floor layout on the contractor, where it belongs, instead of with the architect.

No building or structure can be dimensioned using only one of these three methods. Some partitions should logically be dimensioned to centerline because they align with the centerline of a window mullion. Most partitions should logically be dimensioned to face of wallboard because this is the clear dimension desired and often required.

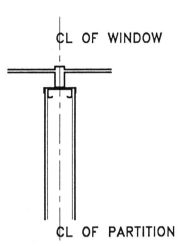

Figure 2.24 Partition located by centerline of window mullion.

Do Not Close Partition Dimension Strings

Partition dimensioning starts with some known position such as an edge of slab, a face of concrete or block wall, or a column centerline. Dimension to all partitions crossing the dimension line, but do *not* close the string to a second fixed-building element. Always leave the dimension string open to allow for building tolerances. Select the least-critical dimension and leave it out. Do not enter an assumed dimension or a "plus-or-minus" dimension. Just leave it out. When dimensional problems come up in the field, it won't be because the architect couldn't add the dimension string correctly.

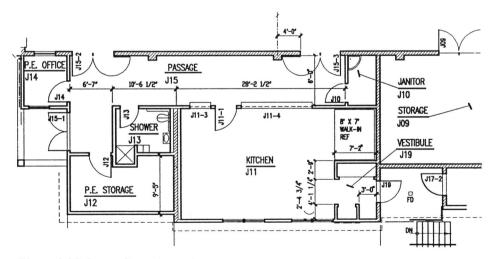

Figure 2.25 Leave dimension strings open.

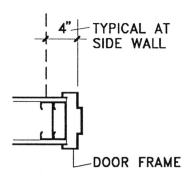

TYPICAL DETAIL

Figure 2.26 Locate door opening by typical jamb detail.

Dimension Doors and Other Wall Openings

Openings in interior partitions are located next. A typical door jamb need not be dimensioned on the plans if a typical detail is drawn to show this condition. Consider four inches to six inches as normal minimum dimensions from hinge jamb to nearest perpendicular side wall. Check this number with the hardware supplier to verify accuracy. Other openings can also be located by dimension on the detail, such as roll-up door tracks and sidelights next to door frames. Dimension all other openings on the plan. Note the *Dimension Point* (DP) on the details, so the contractor knows exactly what point the plan dimension represents. For swing doors, always dimension the hinge jamb or opening centerline. Try to avoid dimensioning the strike jamb because the opening size might change, and if it does, the hinge jamb and centerline dimensions are more likely to remain unchanged. This could save you the cost of a door during construction.

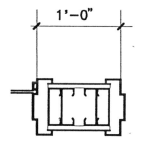

TYPICAL DETAIL

Figure 2.27 Locate sidelight by typical detail.

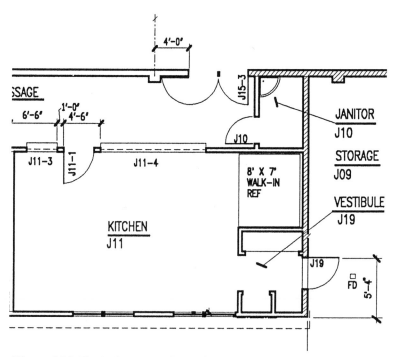

Figure 2.28 Typical opening dimensions in plan view.

Locate Equipment

Specialties and equipment should be dimensioned on detail plans (see drawing scale selection chart, Fig. 2.03.) so they do not clutter smaller-scale drawings. Dimension plumbing fixtures to centerline from surface of room walls. Dimension toilet partitions to surface. This satisfies handicap rules for minimum clearance. Include toilet-partition thickness in dimension string.

Other items which need dimensioning on the detail plan include rows of casework such as in a laboratory, fixed seating in assembly rooms, and fixed equipment. Each dimension must originate at a known reference point such as a wall surface or column grid.

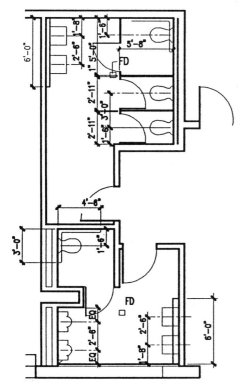

Figure 2.29 Toilet room dimensions.

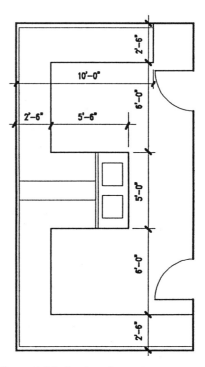

Figure 2.30 Seating dimensions.

Rooftop Dimensioning

Rooftop hatches and equipment also need dimensioning. Select a fixed wall, grid line, or in some cases, a structural member as the reference point. Locate objects by two perpendicular dimensions, and show them on the roof plan. Be sure to coordinate items like skylights which show up on the reflected-ceiling plan.

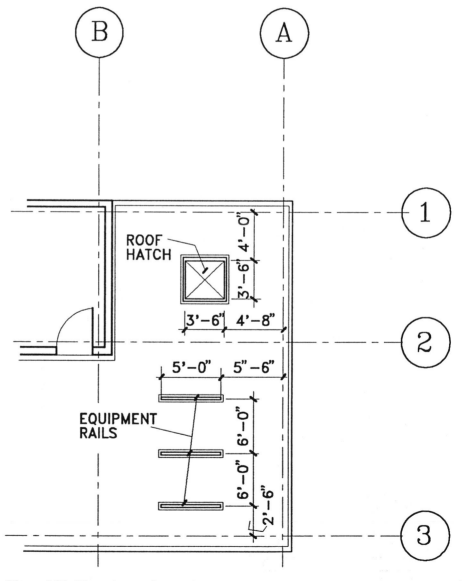

Figure 2.31 *Dimension rooftop equipment.*

Ceiling-Plan Dimensioning

A reflected-ceiling plan usually requires very few dimensions. Grid patterns for suspended acoustical tile usually begin at room midpoint and extend so no border tile is less than half-size and opposite border tiles are equal in size. Occasionally, more control is needed than that, and dimensions are required. Locate ceiling to centerline of suspended grid member from fixed surfaces like room walls. Dimension soffits and borders. Locate light fixtures and mechanical items graphically unless greater control is needed. Dimension skylight wells to the same reference as was used on the roof plan.

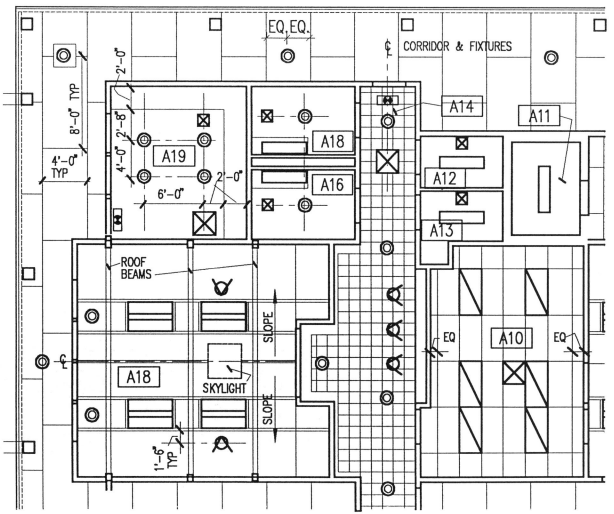

Figure 2.32 *Dimension a reflected ceiling plan.*

Vertical Dimensioning

Building elevations are the first drawings which show vertical distance. Reference floor lines and reference roof lines are extended beyond the building elevation and their relative elevations noted. Like floor plans, a series of dimension lines is placed outside the drawing area. The first line shows overall building height. Dimension from a fixed reference line such as the reference floor line. Never dimension to grade. It is not accurate. This reference top of wall might not be the actual top of coping or fascia, but some internal member, which is easier to control. The detail drawing must show this reference dimension point and any difference to actual top of coping or fascia.

The second dimensions locate wall breaks such as story steps and penthouses. A third line is used to locate reference floor lines, and a fourth line establishes wall opening dimensions.

All dimensions must ultimately be referenced on the construction details which show dimension points as the control for installation and construction.

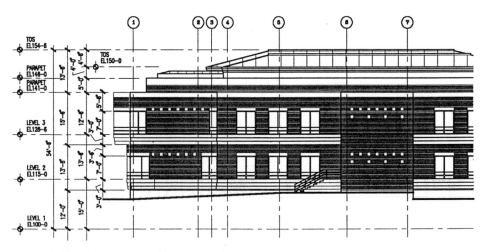

Figure 2.33 *Dimension a building elevation.*

Building-Sections Dimensions

Building sections repeat the four basic exterior dimension strings. They also show changes in floor level within the building and other massing dimensions such as typical ceiling height and special ceiling reference points as in an auditorium.

Building elevations and sections should also show column grids and reference building lines, but no dimensions are required.

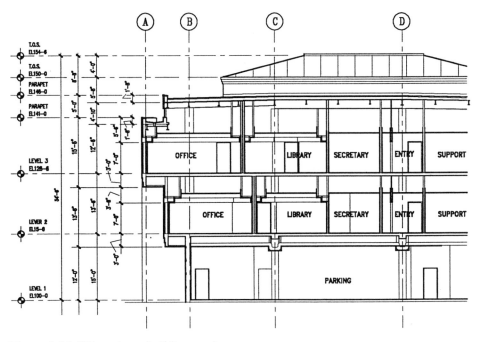

Figure 2.34 *Dimension a building section.*

Wall-Section Dimensions

Wall sections repeat the four basic exterior dimension lines. This may seem to contradict the rule of showing things only once. However, in buildings with complex facades, it is often difficult to show all vertical dimensions on the building elevations and sections. Since wall sections are needed for other reasons (such as detail call out), they are also a good place to show needed vertical dimensions.

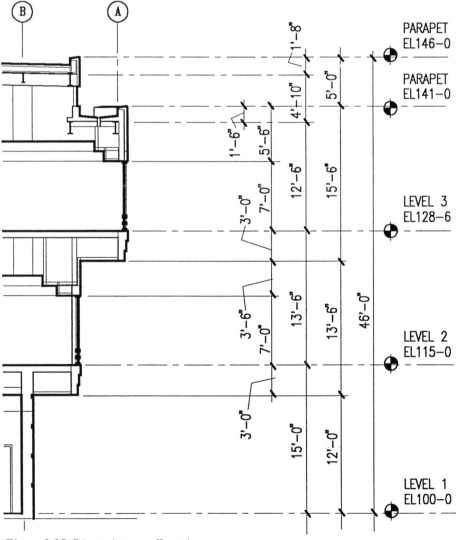

Figure 2.35 *Dimension a wall section.*

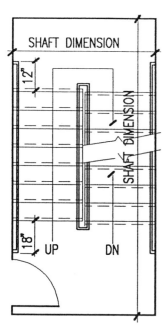

Figure 2.36 *Stair dimensioning in plan view.*

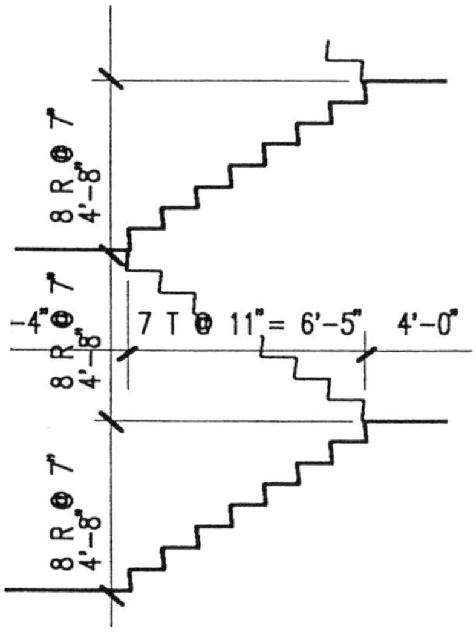

Stair Dimensioning

With the dimension system explained so far, horizontal measure is controlled on the plans, and vertical measure is controlled on the elevations and sections. This rule follows typically for small-scale drawings. One exception is the stair.

A stair is defined in two drawing sets: plans at each floor (or major landing), and sections. The *floor plan* (or *landing plan*) is used to locate the finished shaft by dimensioning the opening to fixed wall or column grid. The stairs themselves are dimensioned only on the section views. All dimensions must close. Plan views are also used to dimension handrail extensions. The horizontal dimension string starts with one wall surface, to the nosing of the first tread, to the nosing of the last tread, and on to the other wall. Always dimension to the nosing of a tread, not the toe of a riser. This will give an accurate description of landing sizes, and locate the control point for handrails.

Vertical dimensions start at reference floor line, to each reference landing line, and close at the next floor line. If the flooring is $1/4''$-thick or less, reference floor line is generally ignored as a problem in maintaining uniform riser heights. However, if reference floor line is depressed from finished floor line by more than $1/4''$, and the stair landing is not depressed, the vertical dimensions must be taken to finished floor. The point here is, riser size must not vary by more than $1/4''$ within a flight of stairs, so choose the dimension point carefully. Stair sections are also used to locate handrails centerline vertically above the stair nosing.

Figure 2.37 *Stair dimensioning in section view.*

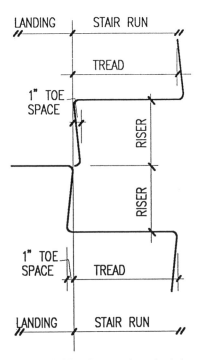

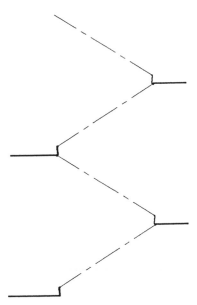

Figure 2.38 *Align nosings in tight stair shafts.*

Some Tips for Dimensioning Stairs

At tight stair shafts, tread nosings should align. The centerline of the handrails will extend into the landing by $\frac{1}{2}$ a tread size plus $\frac{1}{2}$ a railing thickness. This will yield a minimum landing dimension.

If a stair shaft is simple, do not draw a section looking into the risers.

In tall buildings with repeated stairs, show only the top and bottom runs and a line to represent landings and nosings of intermediate runs.

Figure 2.39 *Simplify stair drawing.*

Casework Dimensioning

Dimensions on interior elevations are used to fix items which are a part of and are mounted on the walls. An example is casework. Here, like stairs, both horizontal and vertical dimensions are given. Horizontal dimensions start with the wall surface and size each cabinet, knee space, and equipment alcove. A "scribe" dimension starts each string, to build in a tolerance for walls out-of-plumb and knob projections. All equipment alcoves should be noted with the instruction "HOLD" so everyone knows this space must not get smaller. It can be embarrassing when the dishwasher or other major item arrives and won't fit into the planned recess because someone took up the slack at that location.

Vertical cabinet dimensions are extended outside the field of the casework. Remember to add scribes at top cabinet if you expect a tall cabinet to reach from floor to ceiling.

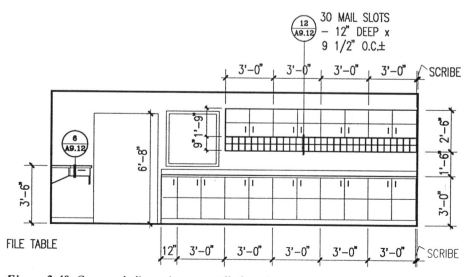

Figure 2.40 Casework dimensions on wall elevations.

Dimensioning Miscellaneous Wall Accessories

Toilet-room accessories are located on interior wall elevations but may more easily be located on a schedule of relationships. This schedule need show these dimensions only once and not for every place they occur. Dimension objects like paper towel dispensers to the outlet opening. This satisfies handicap codes which are concerned about the level a person seated in a wheelchair can reach. Dimension to only one vertical point on any accessory. The final size of the object can vary greatly from the one used for design purposes because owners often subscribe for paper products with vendors who supply their own dispensers. Two dimensions, such as top and bottom from floor, will result in either or both dimensions being wrong.

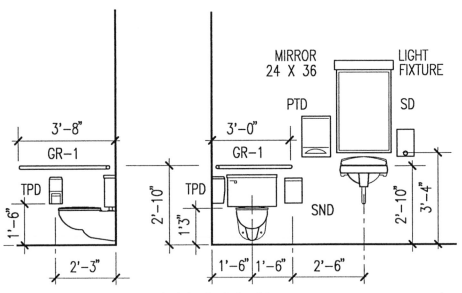

Figure 2.41 *Toilet accessory schedule of relationships.*

The rule of dimensioning to only one point works for most wall-mounted objects. For example, a chalkboard with chalk tray is nominally 4-feet-high, but there are many cases where this will not be the actual size. In most cases, it doesn't matter. One dimension to top of trim at head will align the board with adjacent door frames. A single dimension to chalk tray will fix the board from that reference point. Again, two dimensions will result in errors, and the chalkboard may not be installed where it is really wanted.

Detail Dimensioning

Details represent a variety of problems in dimensioning. Here are a couple of quick rules to consider:

Both horizontal and vertical dimensions can occur on details depending on the subject. Do not dimension standard manufactured items like a wood 2-×-4 or a 4-×-4-×-¼″ steel angle.

Dimensioning starts with a known point such as grid line, floor line, face of building line, or dimension point. Each detail should have both horizontal and vertical reference lines or dimension to one of each.

Fully dimension the detail object or assembly to describe its contents.

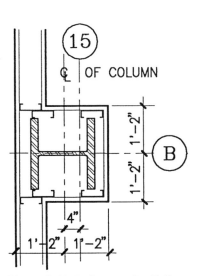

Figure 2.42 Reference detail dimensions to basic building reference lines.

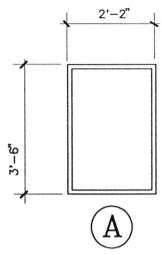

Figure 2.43 *Window type with insufficient information.*

Philosophy of Detail Development

So far, this chapter has been dealing with the importance of drawing scale and a method of locating items in a building by dimensions. Following is a discussion of the philosophy of drawing development.

A drawing doesn't just happen, at least not a successful drawing. It begins with planning. In many ways, the planning of a drawing problem is very similar to planning the building itself. For example, let's go back to the window in Fig. 2.43. The drawing does not show whether the window is made out of wood or aluminum or some hybrid combination like vinyl-clad wood. The drawing also does not show whether the window operates. It appears to be fixed, but that is not made clear. Information about the finish applied to the sash and frame is also missing. When a drawing is as vague as this one, a lot of the required information is not there. This is an example of putting the cart before the horse. Drawing should be stopped, and additional information gathered before any more drafting is attempted.

The first element in problem-solving is to identify the problem. In the case of this window, the problem can be written out as a list of requirements which the window is expected to meet. A good reference is the American Architectural Manufacturers Association (AAMA) manual on window selection criteria.

WINDOW REQUIREMENTS

U-value of .05

Maximum size = 3-2 by 4-6

Thermal break

Window head to align w/ door head

Max. cost—$24/sq ft

Zero maintenance

No operable sash

Owner doesn't like tinted glass

Figure 2.44 *Identify the problem.*

Investigate Options

Once the requirements have been listed, an investigation of windows which meet the requirements can begin. It may be sufficient to open 10 window manufacturers' catalogs and list the windows that meet your requirements. Check the window configurations offered. Notice special sizes. Read the specifications.

From your list, find at least three manufacturers that seem to meet your needs and call representatives to bring additional information. Consider manufacturers who have given you a good product on previous projects. Ask them to bring full-size samples of a window similar to the one you need along with a full range of finish options and suggested installation details. Some vendors will have their details on CAD so they can be imported directly into your details.

Ask a local glass manufacturer to come in to help you select the proper glass for your needs. Call in the mechanical engineer doing the air-conditioning design. Together, you can make decisions about total glass area for building, glass reflectance, and much more. Assemble all your information. Compare technical specifications, cost, and lead time against your list of requirements. Examine each product in a building which has been occupied for at least one year to determine installed performance. Check manufacturers' references. Review test reports and industry standards.

Marvin

Andersen

Pella

Efco

Figure 2.45 *List window manufacturers and series that meet requirements.*

1" Insulating Glass

Lo "E"

Heat Tempered

Laminated

Float / Plate

Figure 2.46 *Collect glass information.*

SPECIFICATIONS

Figure 2.47 *Obtain specifications.*

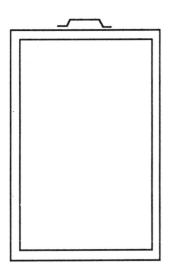

Figure 2.48 *Obtain window samples.*

Update Problem Statement

From this exercise, the list of requirements for your windows can be corrected and expanded. This will become the basis for the specifications for both glazing (CSI section 08800) and windows (CSI sections 08XXX). From the suggested installation details, your own detail drawing can be started.

Select a System

Select one manufacturer as the basis for your window design. If bidding and conditions require, be sure at least two other manufacturers can meet the same parameters as the one you selected.

For this window example, assume the building structure and exterior wall have already been established. The building is steel frame using brick veneer over metal studs as an accent with stucco being the primary finish.

WINDOW REQUIREMENTS

U-value of .Ø5

Maximum size = 3-2 by 4-6

Thermal break

Window head to align w/ door head

Max. cost ~~$24/sq ft~~ $30/sq ft

~~Zero maintenance~~

~~No operable sash~~ OK

~~Owner doesn't like tinted glass~~ OK

No painted wood sash

Figure 2.49 *Update problem statement and technical data.*

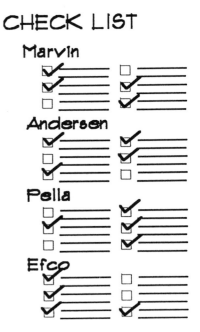

CHECK LIST

Marvin

Andersen

Pella

Efco

Figure 2.50 *Select a system.*

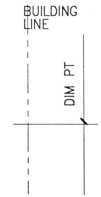

Figure 2.51 *Establish reference grids.*

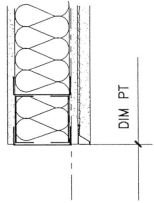

Figure 2.52 *Show wall members.*

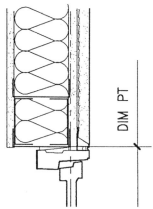

Figure 2.53 *Add window and trim.*

Locate the Detail Assembly in Two Dimensions

As established under dimensioning, every drawing must have two crossing reference lines. These reference lines locate the drawing's contents within the building. Since this basic concept is so important, these reference lines should be the first things drawn.

The vertical reference in this detail is reference face of building. For this building type, the face of exterior sheathing makes a good location because it is common to both the brick and the stucco. The horizontal reference is the reference floor line. However, in a detail, like a window head or sill, this reference line would not fit on the drawing, especially at 3″ = 1′-0″ scale. It is important, then, to make reference by some other method. Use the dimensions shown on the building elevations and wall sections. They, in turn, are shown relative to the reference floor line.

So there are now two reference lines. Begin the detail with the known elements. Locate the back side of exterior sheathing and stud thickness. Determine the header condition for the size of opening. Ask the structural engineer for assistance. Don't rely on tables from manufacturers' data. Determine shim spacing and position the window frame with desired setback from reference building line. Many windows come with a "nailing flange" which automatically positions the window against the sheathing. Add exterior and interior finishes. The basic picture is almost finished at this point.

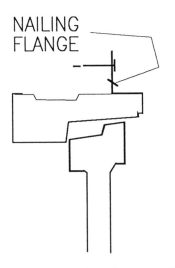

Figure 2.54 *Nailing flange attachment.*

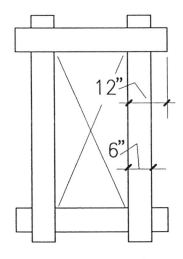

Figure 2.55 *Flash around window.*

Check Anchors and Fasteners

At this stage of detail development, check for two important items: attachment of the window and other elements of the detail, and the ability of the detail to keep out water.

The window described has a nailing flange. For simple structures, the window manufacturer can recommend anchoring devices and spacing. They usually suggest that you not attach the head to the framing so the wall can deflect independently from the window. If the building is subject to excessive wind conditions, have the structural engineer design window anchors. You don't want to be the first architect in town to have a window blow out in a high wind.

Check for Water Integrity

Next, evaluate the detail's ability to keep water out. Windows are notorious for leaking, not at the sash, but around trim and shim joints. To avoid this, use 6″-wide strips of 15-lb. asphalt-saturated felts as a flashing around the nailing flange. Place the sill strip first, then side strips, and finally the head strip. Lap each about 12″. This is very important to water integrity, and must be shown in detail. The head detail in this example is not the place to show how frame flashing is overlapped. Draw a separate elevation detail showing just the flashing. Next show sealant along joints in finish materials and the window frame. Don't assume that two objects shown tight together can keep out water.

Add Notes to Clarify Materials and Give Instructions

The detail is now ready for notes and dimensions. The dimensions should extend to the far right side of the detail space so they are out of the way for note leaders. Show the rough-opening and window-size dimensions using the same dimension points shown on the window-type schedule. If the window-frame face dimension is critical to design, show it. That way, the other acceptable bidders know that this is a requirement. Show shim space dimension, especially if it is expressed as an architectural element. Dimension everything that is critical, otherwise the window manufacturer and contractor might interpret the drawings differently. Ambiguity in dimensioning can lead to misalignment of important architectural elements.

The next step involves placing notes and other references. Chapter 1 discusses graphic symbols and detail-specific notes. When using these standards, remember to check back with the specification to be sure you are using the same language. Note on the detail all materials shown, even those with poché. Note item as "continuous and discontinuous." Do not use terms like "typical" unless they are.

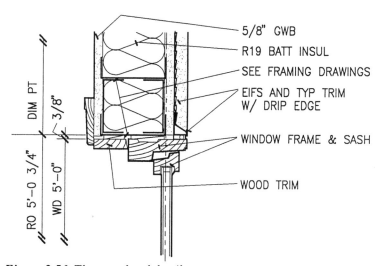

Figure 2.56 *The completed detail.*

CONCLUSION

Detail Development and Formulation

The process described to solve the window problem can be used for any detail drawing problem. The steps, once again, are:

- ### Identify the problem.

- ### Investigate options.

- ### Evaluate the solutions and select one to use for the project.

- ### Select the appropriate drawing scale.

- ### Build up a detail going from the known elements toward the unknown elements. Solve problems as you go.

- ### Locate the detail assembly in two dimensions.

- ### Check anchors and fasteners.

- ### Check for water integrity.

- ### Add notes to clarify materials and give instructions.

- ### Add other references to additional details.

Of these basic drawing elements, dimensioning is of special importance because it locates a drawing item within a building.

ANATOMY OF A SCHEDULE

3

INTRODUCTION

A *schedule* is a concentration of information about a specific subject in either picture or matrix form. A *Door Type Schedule* showing door slabs and vision panels is a good example of a *picture schedule*. A *Room Finish Schedule* with columns and rows of boxes is a *matrix schedule*. Each one represents a lot of information in a rather small space, and this is its primary function. When there is a one-to-one ratio of information, the advantage of preparing a schedule is clearly the graphic presentation which, when done properly, can make the presentation of comparative information both concise and very readable. There are many different schedules and forms used in architectural-contract-document production. This chapter will cover them one by one with examples of how to prepare them and why they are used.

The List

The most simple form of matrix schedule is a *list*. Every drawing set starts with one: *the Index to Drawings.* This is an example of a simple matrix schedule with a left-hand column of information (the drawing-sheet numbers in the *Y* axis) compared to a horizontal header (the drawing-sheet title in the *X* axis). In a list schedule like this, there can be only one column of information for each column of sheet numbers.

The Matrix

The most typical architectural matrix schedule, whether produced by manual drafting techniques or by *Computer Aided Production* (CAP), is a simple array of boxes which contains the desired information. The schedule title is usually on the top followed by a row of column headings. The left-hand column (*Y* axis) will contain a list of items to be amplified or compared to all other columns. Lines are drawn to form a matrix where information is added such as a check mark to denote a connection between two items.

When designing a matrix schedule, attention should be given to letter point size and density and to line weight. (Refer to Chap. 1 for basics on graphic standards). The minimum lettering height may be dictated by the local building department as $\frac{1}{8}''$-high. For hand lettering, use a soft pencil which will produce a dense letter. For automated processes, choose a regular dense letter. Fine and bold lettering should be used for accents. Often a schedule will read best when there is a heavy line separating the heading from the main content and another heavy line separating the left column from the main contents. Another graphic method is to have horizontal lines every five rows through the field rather than every row. With computerized schedules like Lotus 123™ and Excel™, many other graphic elements can be added to make a schedule more readable.

INDEX TO DRAWINGS

- A0   DRAWING INDEX
- A1   SITE PLAN
- A2   FLOOR PLAN
- A3   BUILDING ELEVATIONS
- A4   BUILDING SECTIONS
- A5   WALL SECTIONS
- A6   SCHEDULES
- A7   INTERIOR DETAILS

Figure 3.01 *The Index to Drawings is an example of a list schedule.*

ROOM		FLOOR
NO.	NAME	MAT/FIN
100	LOBBY	STONE/
		FLAME
100A	RECEPTION	STONE/
		FLAME
		CARPET/
		FF
101	CORRIDOR	CARPET/
		FF
102	OFFICE	CARPET/
		FF
102A	CLOSET	CARPET/
		FF

Figure 3.02 *The Room Finish Schedule is an example of a matrix schedule.*

Forms

Forms provide a method for gathering and illustrating information. Unlike schedules, they are more random in appearance, with lists and more than one set of *X-Y* relationships. When forms are used to gather information, they contain blank spaces to be filled in as the information is obtained. Forms can also be used to disseminate information, as in transmittal forms. This form is really a combination of form and schedule, since it contains both fill-in spaces for destination and a matrix schedule for a list of transmitted data.

Every office form should contain the form's name and the name, address, phone, and fax numbers of the design professional's business. Consider a well-prepared form as a type of advertising. You want the users to see your name on it.

Room Data Sheet

One of the first forms to prepare on any project is the *Room Data Sheet*. This form will be used to gather information which is later used to prepare the Room Finish Schedule, Door and Opening Schedule, and other contract documents. It contains space to note room size, length and width, finishes, ceiling height, door openings and types, casework, and general equipment needs. It also has space for plumbing, air-conditioning, electrical, lighting, and communications needs.

1 Identification:
The first area of the form is for recording project title, date, job number, and the name, phone, and fax of the person who filled out the form. Next is space for department, room name and number, and room size in length, width, height, and area.

These entries, right up front, identify the main purpose for the form where it can be found quickly.

BASSETTI NORTON METLER REKEVICS ARCHITECTS

Page 1 of 2

ROOM DATA SHEET
BASSETTI NORTON METLER REKEVICS

IDENTIFICATION

Project
Name

Job No.

Date

Prepared By

Phone   FAX

Location
Department

Room Name

Room Numbers

Room Size/
Length   Width   Height   Area

1

GENERAL

Function
General Requirements

Acoustics
Floor

Walls

Ceiling

Others

Occupancy
Classification

Separation

Others

2

FINISHES

FLOOR MAT/FIN	BASE MAT/FIN	WALL MAT/FIN	WAIN MAT/FIN	TRIM MAT/FIN	CEILING MAT/FIN	HT	REMARKS

3

OPENINGS

OPENING QUANTITY	SIZE	RATINGS FIRE	ACOUST	DOOR MAT/FIN	FRAME MAT/FIN	SILL MAT/FIN	HARDWARE REQUIREMENTS	REMARKS

4

ACCESSORIES

Specialties
Visual Display

Toilet & Bath

Floor & Wall Systems

Signage

Casework
Wood   PLAM   Metal   Others

Base Units

Wall Units

Tall Units

Counter Tops

General

5

Figure 3.03 *The Room Data Form.*

2 General:
The second area on the Room Data Sheet is for general architectural design requirements, and a few specifics about acoustics and occupancy. In the general requirements space, write a short narrative of the intended room function. Note acoustic ratings required for demising walls, floors, and overhead structures. Fill in the occupancy classification of the room and any anticipated separations required from adjacent spaces.

3 Finishes:
The third area of the form is for finishes. This is a schedule-type form which closely resembles the actual Room Finish Schedule, prompting for information on material and finish for floor, base, wall, wainscot, trim, and ceiling. There is also a space to enter the ceiling height.

4 Openings:
Openings is for doors, relites and other wall openings, excluding windows and louvers. This section of the form is also in schedule format and closely resembles the final opening schedule. Enter information about size and rating. Show material and expected finish for doors, frames, and sills (thresholds). Note hardware requirements such as locks and closers.

5 Accessories:
The fifth area is for miscellaneous specialties and casework. Follow the prompts as they ask for information. Visual display is for chalkboards, tack boards and marker boards. Toilet and bath is for any bath or hand-wash accessories. Floor and wall systems refers to items like computer access floors and demountable walls and partitions.

The casework area has space to identify cabinet material. Next, enter descriptions for each kind of cabinet and counter tops, such as open shelving and sink base.

This concludes side one of the Room Data Sheet. Side two continues the questionnaire.

Furnishings
Furniture

_____
_____
_____
_____

Window Treatment

Seating

General

_____
_____
_____

Equipment
Fixed Equipment

_____
_____
_____

Major Moveable Equipment

_____
_____
_____

Minor Moveable Equipment

_____
_____

6

Plumbing
Sink(s)

| CW | HW | Others |

Standard Waste   Plaster Trap

Floor Drain   Floor Sink

Oil Sepapator   Acid Resistive Waste

Urinals   WC

Additional
Requirements

HVAC   FLOOR
Heating   Cooling

Exhaust   Air Changes

Room Pressure   Positive   Negative

Min. Temp.   Max. Temp.

Relative Humidity   Filtering

Control

Additional
Requirements

7

Electrical
Normal Power   120V   208V   Others

Emergency Power 120V   208V   Others

Dedicated Power   Clean Power

Additional
Requirements

Lighting   General

Task

Additional
Requirements

Signal & Communications
Telephone

Intercom

Computer Terminal

Building Security

Additional
Requirements

Comments

8

ROOM DATA SHEET
BASSETTI NORTON METLER REKEVICS

FURNISHING & EQUIPMENT

SYSTEMS

GENERAL

BASSETTI
NORTON
METLER
REKEVICS
ARCHITECTS

Page 2 of 2

Figure 3.04 _Room Data Form, side two._

6   Furnishings and Equipment

Side two of the Room Data Sheet begins with furnishings and equipment. Building furnishings are often provided by the building owner or tenant, but may become a design element for the architect's consideration. Enter items such as office furniture and waiting-room seating.

Equipment is divided into three main groups. *Fixed equipment* is any item which is securely anchored to the building. *Major moveable equipment* is not anchored, but often requires rough-ins for power or mechanical services. *Minor moveable equipment* includes wastebaskets and tissue holders. Additional forms will be used to prepare a more finite description of all equipment.

7   Systems:

Systems include plumbing, HVAC, power, lighting, and communications. The form is divided into areas to address each of these.

Plumbing requirements for water, waste, and fixtures are requested. Add requirements for special traps for plaster sinks and oil separators.

Note requirements for HVAC, including a need for heating or cooling, such as cold rooms and greenhouses. Note exhaust and air-change requirements, as well as room pressurization, temperature, and humidity.

Electrical requirements are divided into two areas: power and lighting. Note basic room-power needs, including dedicated circuits and clean power. Describe both general and task lighting needs.

Signal and communications needs are next. Note requirements for telephones, intercom, and computer terminals. This section is also used to describe room-security needs.

8 <u>General:</u>
The last area of the Room Data Sheet is for general comments. This is the place to enter room requirements that don't seem to fit elsewhere on the form. Information might include a statement about the time of day the room is expected to be occupied. This could effect the mechanical requirements of heating and cooling.

Once the Room Data Sheets have been completed, the next family of forms, including Room Finish Schedule and Door and Opening Schedule, can be started.

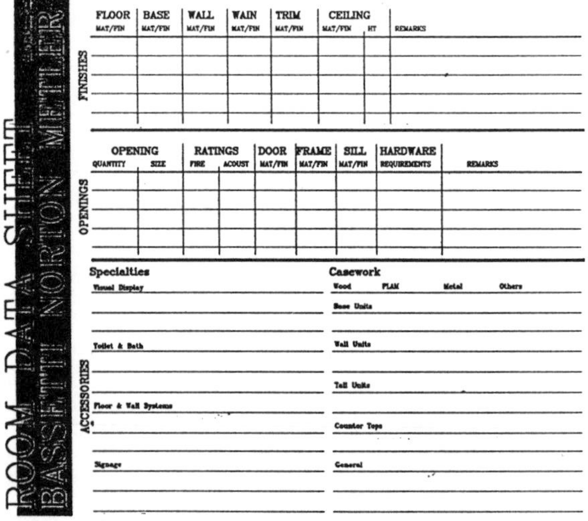

Figure 3.05 *The Room Data Form, side one.*

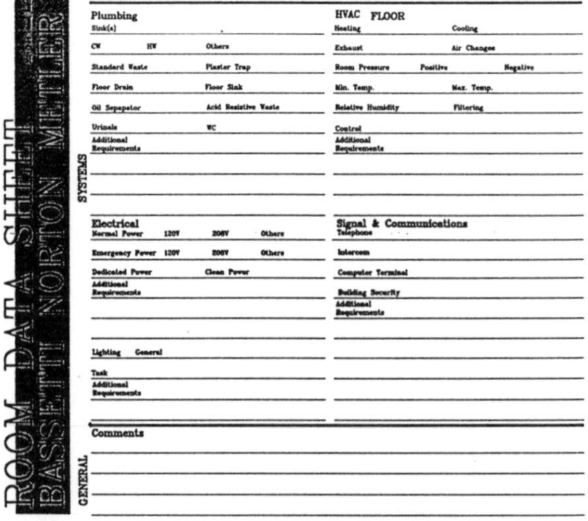

Figure 3.06 *The Room Data Form, side two.*

ROOM FINISH SCHEDULE

Room finishes, even when abbreviated, usually will not fit within a room on an $\frac{1}{8}$"-scale floor plan. One way to solve this problem is to draw interior elevations of every room. These elevations, along with floor and reflected ceiling plans, could provide enough space for noting finishes, but it would take a lot of drawing and a lot of time. Unless the wall elevation was needed for some other purpose, there might be a lot of empty boxes in the interior elevations package. The schedule provides a concise format for presenting a vast amount of finish-related information and is far simpler and less time-consuming. One thing the schedule does not do, however, is to show where one finish stops and another finish begins. Many attempts have been made to make the schedule show this separation, and they simply don't work. The plan and elevation views are still required for this purpose. With this in mind, the schedule can be started.

There are two basic methods for recording room-finish requirements: the *Schedule of Finishes* and the *Room Finish Schedule*. The most frequently used method is the Room Finish Schedule. It starts like other schedules, with the title and field headers on top and a list of room numbers along the left border. The resulting matrix is filled in with room names, followed by finish requirements, using abbreviations from the material/finish abbreviations list and the standard list (see Chap. 1).

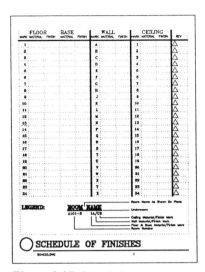

Figure 3.07 *Schedule of Finishes.*

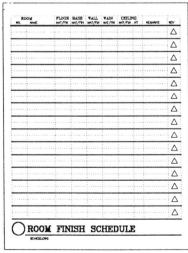

Figure 3.08 *Room Finish Schedule.*

Material/Finish Abbreviations List

Finishes are noted by showing both material and finish. In this way, a wall surface of gypsum wallboard (GWB) can show a finish of flat wall paint (PT-1) or vinyl wall covering (VWC). When materials come finished from the manufacturer, like vinyl composition tile (VCT), they are simply noted as factory finish (FF) unless waxing and polishing are required (W & P). For some materials, like carpet (CARP), no additional finish is required, neither at the factory or in the field. The finish is then entered as a dash (—).

ACT	Acoustical Ceiling Tile
ANOD	Anodized
AWP	Acoustic Wall Panel
CARP	Carpet
CEM PL	Cement Plaster
CMU	Concrete Masonry Units
CONC	Concrete
CT	Ceramic Tile
ENAM	Enamel
ETR	Existing To Remain
FBD	Fiber Board
FF	Factory Finish
FIN	Finish
FWP	Flat Wall Paint
GALV	Galvanized
GFRC	Glass Fiber Reinforced Concrete
GFRP	Glass Fiber Reinforced Plaster
GLAM	Glue Laminated Wood
GWB	Gypsum Wall Board
HCT	Hollow Clay Tile
HDBD	Hardboard
HDWD	Hardwood
HM	Hollow Metal
MCW	Mineral Core Wood
MDF	Medium Density Fiberboard
MET	Metal
OSB	Oriented Strand Board
PLAM	Plastic Laminate
PLAS	Plaster
PLYW	Plywood
PNT/PT	Paint
POL	Polish
QT	Quarry Tile
RB	Rubber Base
S&V	Stain & Varnish
SCW	Solid Core Wood
SFC	Special Floor Coating
SGEN	Semi Gloss Enamel
SHV	Sheet Vinyl
SLR	Sealer
SST	Stainless Steel
SWC	Special Wall Coating
TER	Terrazzo
UNFIN	Unfinished
VCT	Vinyl Composition Tile
VP	Veneer Plaster
VT	Vinyl Tile
VWC	Vinyl Wall Covering
WD	Wood

Figure 3.09 *Material/finish abbreviations list.*

Note Materials and Finishes Accurately

When scheduling materials and finishes, be careful to note them accurately. For example, a concrete floor is a material when it is sealed or painted, but it is not the material when it is covered with sheet vinyl or carpet. Gypsum wallboard is a material when covered by paint or vinyl wall fabric, but it is not the material when it is covered by wood paneling or stone veneer.

ROOM		FLOOR	BASE	WALL	WAIN	CEILING		REMARKS	REV
NO.	NAME	MAT/FIN	MAT/FIN	MAT/FIN	MAT/FIN	MAT/FIN	HT		
									△
									△
									△
									△
									△
									△
									△
									△
									△
									△
									△
									△
									△
									△
									△

◯ ROOM FINISH SCHEDULE

Figure 3.10 *The Room Finish Schedule.*

The Room Finish Schedule is a matrix with headers across the tops of 10 basic columns.

1   Column number 1, Room Number:

Room numbers are listed in alphanumeric order as shown on the floor plans. (Room numbers are derived in a number of ways and are described in greater detail in Chap. 9, "Systems".) For now, let's assume a system of numbering rooms with three digits. The first digit represents the floor or story followed by 01 through 99 consecutively for each room. Rooms which are part of a larger room, like a closet, don't usually rate a new number. They get a letter suffix, i.e., 103A, which quickly relates it to the main room, 103. This number is then entered in the first column on the schedule.

| ROOM | | FLOOR | BASE | WALL | WAIN | CEILING | | | REV |
NO. NAME		MAT/FIN	MAT/FIN	MAT/FIN	MAT/FIN	MAT/FIN	HT	REMARKS	
									△
									△
									△
									△
									△
									△

Figure 3.11 *Room Finish Schedule column headings.*

2   Column number 2, Room Name:
   Enter the room name exactly as it appears on the plans. If it is abbreviated on the plans, abbreviate it here. Claims against the architect have developed because of such seemingly insignificant differences. If you like to write out the whole name in the finish schedule, be sure there is room to write it out on the plans.

3   Columns number 3, 4, 5, 6, and 7:
   These are for entries of material/finish for floor, base, wall, wainscot, and ceiling, respectively.
   Where two or more materials exist, enter them in successive spaces. Their locations must be shown elsewhere. Do not try to write a schedule which attempts to define the boundaries between two different materials on the same surface. If you have a wood floor with carpet inset, this can not be clearly shown here. You must draw the condition on the floor plan. Even the wainscot column does not fully address the condition. An interior elevation is still required to show height and top-edge condition. Enter the material and finish here, and let the other drawings explain where they go.

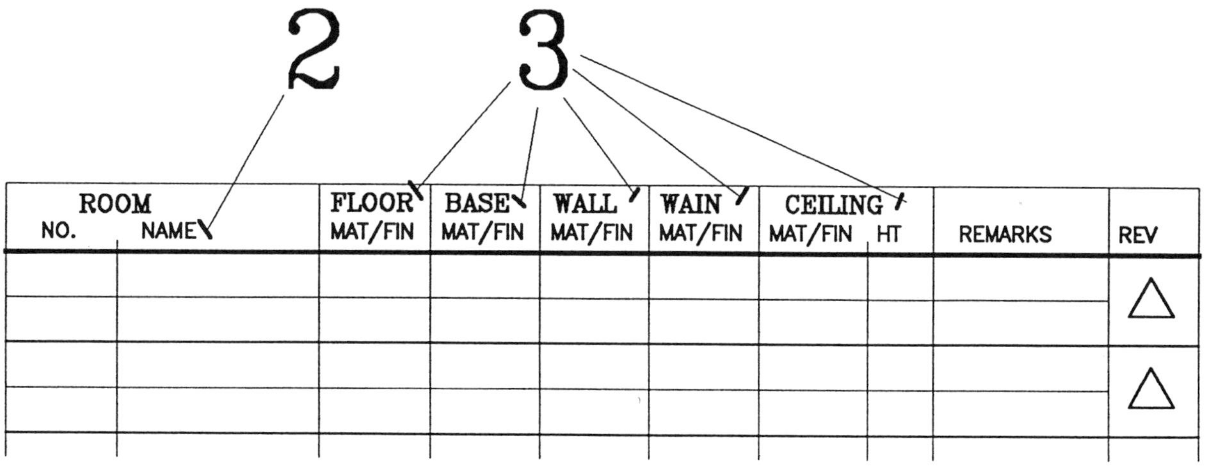

Figure 3.12 *Room Finish Schedule column headings.*

4 Column 8, Ceiling Height:
There has been much discussion about whether to have this column or to note ceiling height on the reflected ceiling plan. This column is here because there are conditions which do not require a reflected ceiling plan. When this happens, this column should be used.

5 Column number 9, Remarks:
Save the remarks column for noting rare conditions which do not require an additional drawing and for referencing a detail such as a special base condition or a floor-transition detail.

6 Column number 10, Revisions:
This is for recording revisions to the finish selections entered in each room. If, for example, the floor and base in rooms 103 and 105 are changed from ceramic tile to sheet vinyl, and this is the first revision to the schedule, place a "1" in the revisions column at each room.

| ROOM | | FLOOR | BASE | WALL | WAIN | CEILING | | REMARKS | REV |
NO.	NAME	MAT/FIN	MAT/FIN	MAT/FIN	MAT/FIN	MAT/FIN	HT		
									△
									△

Figure 3.13 Room Finish Schedule column headings.

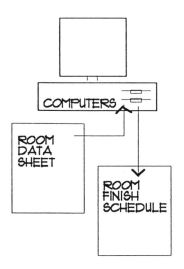

Figure 3.14 *Automate room finishes by linking to data-base file of Room Data Forms.*

Link Room Data and Finish Schedule by Computer

The Room Finish Schedule is a familiar method for displaying the material/finish requirements for a project, and can be derived directly from the Room Data Sheets. When computerized production methods are employed, material/finish requirements can be updated en masse and shared between room data files and room finish files. When this linking of data is employed, the concept of using the Room Finish Schedule becomes very useful. However, on very large projects, this form of schedule can become very lengthy, to the point of taking sheets and sheets of drawings. A more space-efficient way to display room material/finish is the *Schedule of Finishes*.

THE SCHEDULE OF FINISHES

The *Schedule of Finishes* does only one thing: It lists all material/finish selections for a project and relates them to a mark which appears on the floor plan. It generally takes no more space than two standard-size detail spaces 7¼-inches-by-9-inches, so it can fit almost anywhere.

Finishes Are Referenced on Floor Plans

The process starts on the floor plan where room names are entered. Each room name is underscored, followed by two entries: the room number on the left, and material/finish marks on the right. Unlike the Room Finish Schedule, which starts with the room number, the Schedule of Finishes does not use the room number at all. It is as if the room title and number on the floor plan were an extension of the Schedule of Finishes, and consequently, there is no need to repeat them.

Figure 3.15 *Finish marks on a floor plan.*

Making Entries on the Schedule of Finishes

The material/finish marks are listed in three columns on the Schedule of Finishes. The first column of numbers represents the floor and base. The second column of letters represents the walls, and the third column of numbers represents the ceiling. Wainscots are not listed out separately so they must be treated as part of the wall finish. There is no entry for ceiling height, so reflected ceiling plans are required if the Schedule of Finishes is used.

Abbreviating Material and Finish

Materials and finishes are entered using the same rules as for the Room Finish Schedule, except that they are only entered once. In this manner, floor mark "1" becomes "VCT/W & P," wall mark "A" becomes "GWB/PT-1," and ceiling mark "1" becomes "AC-1/FF." Continue down the schedule to enter all material/finish requirements for the project. Then go back to the floor plan and enter the marks for the appropriate material/finish under each room name.

MARK	FLOOR MATERIAL	FINISH	BASE MATERIAL	FINISH	MARK	WALL MATERIAL	FINISH	MARK	CEILING MATERIAL	FINISH	REV
1					A			1			△
2					B			2			△
3					C			3			△
4					D			4			△
5					E			5			△
6					F			6			△
7					G			7			△
8					H			8			△
9					J			9			△
10					K			10			△
11					L			11			△

Figure 3.16 *Sample Schedule of Finishes.*

Benefits and Drawbacks

The benefit of this system is that everyone can tell what finishes are required in a room just by looking at the room name. For simple projects, the marks are quickly memorized, and for marks which are not often used, the schedule can be copied and pasted to the back of the previous drawing sheet for easy reference.

A drawback to this system is that every room must have a material/finish entry. On large jobs this can be very time-consuming, especially if you were to change the floor in all classrooms in a school from VCT/W & P to CONC/SH. Even so, this is generally less work than preparing a Room Finish Schedule by manual means.

Both of these methods of scheduling room-material/finish requirements should be part of an office production manual, but the subtle differences between the two must be understood. As mentioned, the Room Finish Schedule has space to enter ceiling height, and the Schedule of Finishes does not. So, if you intend to use the Schedule of Finishes, it is likely that you will need to draw reflected ceiling plans to show ceiling height.

The major advantage to using the Room Finish Schedule over the Schedule of Finishes comes with computer automation. Start with the Room Data Sheets by entering material/finish information. Link up to the Room Finish Schedule using a data-base manager program. Now, the computer can automatically fill in the Room Finish Schedule. Further, when the classroom changes from VCT/W & P to COMC/SH, it can be done by making one mass update for all rooms named "classroom." When rooms are added, the computer can automatically place the new room in alphanumeric order. In this manner, it is relatively easy to keep the Room Finish Schedule updated.

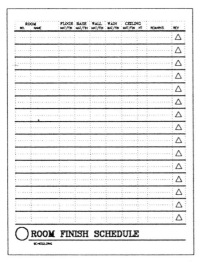

Figure 3.17 Room Finish Schedule.

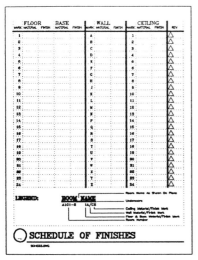

Figure 3.18 Schedule of Finishes.

Recap on Finish Schedules

To recap, the material/finish requirements for a project can be scheduled by one of two methods: the Room Finish Schedule which lists every room, or the Schedule of Finishes which lists only the material/finish combinations. Each has its use, and the decision of which to employ must be weighed relative to the total project.

DOOR AND OPENING SCHEDULE

Requirements for doors, rolling grilles, etc. are taken from the Room Data Sheets where number and size are given along with general requirements for glass and locks. This is the beginning data for the Door and Opening Schedule. It is not, however, by far, the end. Additional data includes door and frame material and finish, glass size and type, installation details, hardware requirements, and more. All of this information could be shown on one schedule, but it would become too large to draw or reproduce. So, it is best to start with one master schedule, the Door and Opening Schedule, and use supporting schedules for door types, frame types, glass types, paint and other finishes, etc.

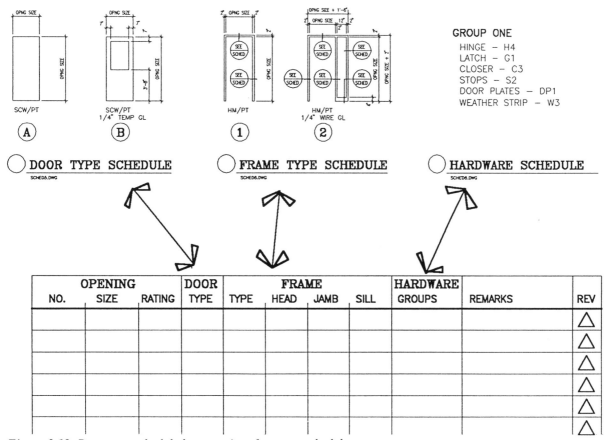

Figure 3.19 *Doors are scheduled on a series of support schedules.*

The Door and Opening Schedule starts with a title and field headers across the top and a list of door numbers down the left-hand column. The resulting matrix is filled in with data describing each door.

1   Column number 1, Opening Number:
Doors are numbered using the number of the room which they "serve." This is important. Don't use the "other" room as the basic door number because that number could be used many other times, i.e., rooms off a corridor. When more than one door exists in a room add a "-1," "-2," etc. after the room number until all doors serving a room are numbered. Number all doors, rolling grilles, roll-up countertops, gates, and framed openings. Number every opening which has a frame or has hardware (except cabinetwork and access panels). Store fronts might be referenced elsewhere, but the doors still need numbers. Also number existing doors that might possibly be altered, even by just painting or rekeying.

2   Column number 2, Opening Size:
For a single door this is for the nominal door size i.e., 3-0-x-7-0 recorded in feet and inches, but without the foot (') and inch (") indicators. For pairs of doors, the opening size is the total nominal dimension across the frame. An example of this is two 3-0-x-7-0 doors in one opening, the opening size is 6-0-x-7-0.

| NO. | OPENING | | DOOR | FRAME | | | | HARDWARE | REMARKS | REV |
	SIZE	RATING	TYPE	TYPE	HEAD	JAMB	SILL	GROUPS		
										△
										△
										△
										△
										△
										△

Figure 3.20 *Door and Opening Schedule column headings.*

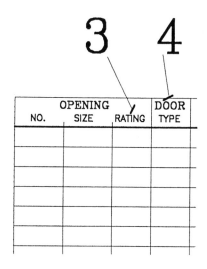

Figure 3.21 *Door and Opening Schedule column headings.*

3 Column number 3, Fire Rating:
This is a number in minutes for 20-, 45-, 60-, and 90-minute openings and hours for 3-hour openings. This designation corresponds to NFPA-80 designations for fire-resistive requirements and labeling of doors and frames. Most manufacturers must follow this same rating system in order to qualify as a bidder.

4 Column number 4, Door-Type Mark:
The door-type mark, always a letter or double letter, comes from a supporting schedule of door-slab elevations. Usually drawn at ¼″ = 1′-0″ or ⅜″ = 1′-0″ (see Chap. 2 on drawing scale), the door-type schedule starts with the most common door configurations of wood, metal, aluminum, and all combinations of panels, vision panels, and louvers. Overall dimensions for door slabs are not given (see "opening size") so the slab drawing can apply to more than one size.

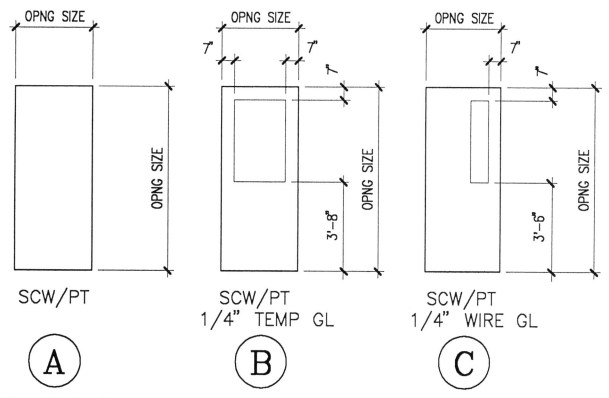

Figure 3.22 *Door Type Schedule.*

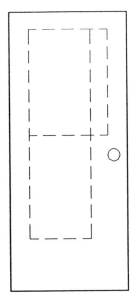

When specific dimensions for stiles and rails and for vision panels are desired, they can be used without closing the string. Vertical dimensions for vision panels must be carefully coordinated with hardware locations so the cutouts leave enough wood at rated conditions. For this reason, it is best to dimension down to the top edge of glass and up to the bottom edge of glass, and leave glass size for shop drawings. Under each slab elevation is the letter indicator identifying the slab, and a note for material/finish. Also noted are door thickness and glass type, if used. When two doors occupy the same opening as a pair, this is noted by the suffix "PR" after the door-type identifier, i.e., "C-PR."

Figure 3.23 Maintain clearance around hardware at fire doors.

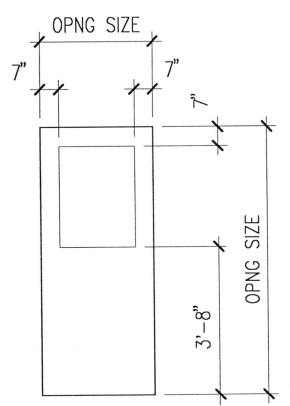
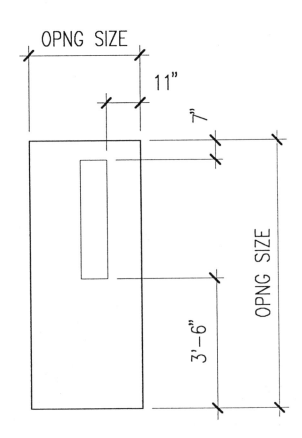

Figure 3.24 Door Type dimensioning.

5 Column number 5, Frame Type Indicator:

This number comes from a supporting schedule of frame types and is usually drawn at $\frac{1}{4}'' = 1'-0''$ or $\frac{3}{8}'' = 1'-0''$ scale (see Chap. 2 on drawing scale) to match the door-type schedule. This Frame Type Schedule is a pictorial representation of each frame including wood, hollow metal, and aluminum. It also is used to illustrate openings for relites, roll-up grilles, and combinations of doors, grilles, side lights, etc. The Frame Type Schedule is not used to illustrate windows and storefronts (see Window Schedule later in this chapter). Frames are drawn and dimensioned with emphasis on "opening size," "frame size," and "rough opening." *Opening size* is used whenever a door or pair of doors is present in the frame. *Frame size* represents the overall frame dimension, and *rough opening* is the required wall opening used in masonry and concrete construction.

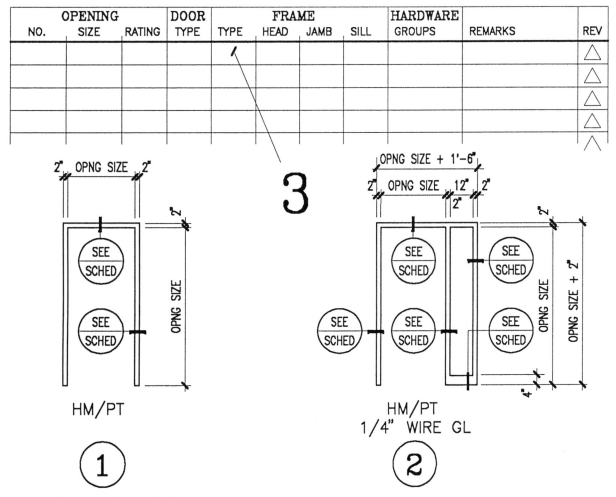

Figure 3.25 *Frame Type Schedule.*

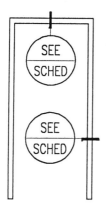

The *frame elevation* is also used to make reference to installation details. For simple four-sided openings with only one reference each for head, jamb, and sill, the detail reference says, "SEE SCHED," referring to the columns in the Door and Opening Schedule. In this manner, a single-frame elevation can work for many installation conditions. For more complex frames, the complete detail references are made on the frame elevation. The Door and Opening Schedule entries for head, jamb, and sill receive an entry referring to numbered notes in the remarks column.

When two frames are exactly alike, but the installation details are not, a new frame type should be drawn to avoid confusion. When two frames are exactly alike except for a minor difference like glass type, the same elevation may be used with an additional frame-type indicator next to the first title, and a note explaining the difference.

Figure 3.26 *Detail references for simple frame types.*

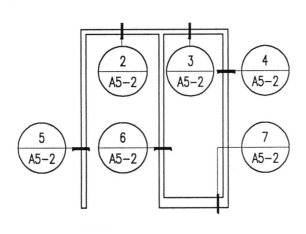

Figure 3.27 *This is an example of a detail reference for complex frames.*

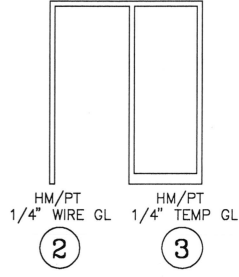

Figure 3.28 *Note minor differences in frame conditions.*

The next three columns are used to identify installation details for simple, four-sided openings as mentioned under frame-type above.

6 Column number 6, Head:
This column is used to reference the opening-head condition. If the opening requires more than one head-installation detail such as a door with adjacent side light in the same frame, the head details should be referenced on the frame type.

7 Column number 7, Jamb:
This column is used to reference the opening jamb conditions. For most openings, both jambs are exactly alike, and only one entry is made in this column; however, for some openings, each jamb has its own condition. For these situations, simply enter both detail references in the jamb column. The frame type does not indicate "handing," and the conditions are obviously related to one or the other jamb by looking at the floor plan. No further reference is needed to tell the contractor which jamb receives which installation detail.

NO.	OPENING SIZE	RATING	DOOR TYPE	TYPE	FRAME HEAD	JAMB	SILL	HARDWARE GROUPS	REMARKS	REV
										△
										△
										△
										△
										△
										△

Figure 3.29 Door and Opening Schedule column headings.

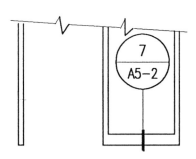

Figure 3.30 *Sill reference for complex frames.*

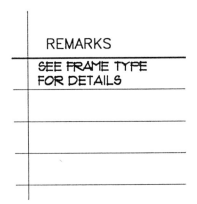

Figure 3.31 *Schedule notes when details are referenced on frame types.*

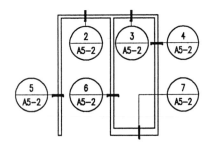

Figure 3.32 *Frame elevation with detail references.*

8 <u>Column number 8, Sill:</u>
This column is used to record threshold conditions for door openings and sill conditions for solitary relites, louvers, and roll-up grilles. If the opening requires more than one sill-installation detail, such as a door with adjacent side light in the same frame, the threshold and sill references should be made on the frame type.

The referencing conditions described for head, jamb, and sill should be followed as closely as practical; however, on very large projects, the number of frame types could become unmanageable if individual conditions are drawn for every installation condition. In this case, entries for head, jamb, and sill can be used for referencing all installation details. The frame type will simply have reference symbols with "See Sched" entered in each location where a detail is required. It becomes obvious which detail goes where by looking at the frame type and the floor plans. However, do not use this method as a short cut for bugging frame elevations on more simple projects. It is better to identify the installation conditions on the frame type so that the contractor does not need to assume anything. When the Door and Opening Schedule is used to call out all installation details, the specifications should require the contractor to produce shop drawings showing every frame with its own installation details. This practice will help reduce errors based on assumed information, and aid in shop-drawing checking.

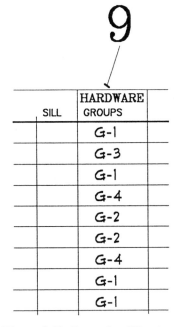

	SILL	HARDWARE GROUPS	
		G-1	
		G-3	
		G-1	
		G-4	
		G-2	
		G-2	
		G-4	
		G-1	
		G-1	

Figure 3.33 *Example of Hardware Groups schedule.*

9 Column number 9, Hardware:

This column is also associated with a supporting schedule. The *Hardware Schedule* is really a list of specific hardware items arranged into groups. Each group lists the following hardware: hinge, lock-latch set, closer, stops, door plate, threshold, and weatherstripping. Some other items less commonly used are electronic strikes and removable center mullions. Since many doors have the same combination of these items, they are each given the same group I.D. on the Door and Opening Schedule, rather than listing each individually. This Hardware Schedule is often prepared by a local hardware supply house or outside consultant. In either case, to make shop-drawing checking easier, be sure the specification requires the successful hardware bidder to submit their hardware-shop drawings using the same format as your consultant has used while writing the Hardware Schedule.

For very large projects, a separate Hardware Schedule of groups becomes unmanageable. For many people, by the time you get to group 20, it is difficult to be sure the next group does not repeat the contents of a previous group. In this case, it would be better to list hardware on the Door and Opening Schedule. For this purpose, the following columns are reserved:

Column 9-H for Hinge
Column 10-G for Lock-Latch Group
Column 11-C for Closer
Column 12-S for Stop
Column 13-D for Door Plate
Column 14-T for Threshold
Column 15-W for Weatherstripping

Entries in each of these columns represents a specific hardware item which is described in the specifications. The advantage of listing hardware requirements directly on the Door and Opening Schedule, beyond the benefit of managing the problem for large projects, is that all the elements of the door-specific hardware are listed after each door number, not on another supporting schedule. If the hardware submittal is organized this same way, it becomes very easy to compare documents.

HARDWARE						
H	G	C	S	DP	T	W
H-3	L-8	MP	S-1	-	T-4	W-1
H-2	L-1	-	S-1	DP-2	-	-
H-1	L-3	MP	S-4	-	-	-
H-1	L-4	-	S-1	-	-	W-1
H-3	L-1	-	S-1	DP-2	-	W-1
H-5	L-8	MP	S-1	-	T-4	W-3
H-1	L-4	-	S-1	-	-	W-1
H-2	L-1	-	S-1	DP-2	-	-
H-1	L-3	MP	S-4	-	-	-
H-1	L-4	-	S-1	-	-	W-1
H-3	L-1	-	S-1	DP-2	-	W-1
H-5	L-8	MP	S-1	-	T-4	W-3

Figure 3.34 *Example of Door and Opening Schedule with all hardware shown.*

10 Column 10 (short form) or Column 16 (long form), Remarks: These remarks should be numbered and referenced back in the previous columns.

11 Column 11 (short form) or Column 17 (long form), Revisions: This column is used to record the revision number for all changes made to the Opening Schedule. If, for example, the jamb-detail reference for opening 103 was incorrectly entered as 3-2, when it is corrected to 3-3, a number is entered in the revision column to indicate the change (see Chap. 8 for more on revisions).

To recap, the Door and Opening Schedule is a master schedule with a series of supporting schedules used to list all doors, relites, framed openings, and other wall openings on a project which require hardware. It can take two basic forms: one for small projects with grouped hardware, and one for larger projects with over 20 hardware groups.

	OPENING		DOOR	FRAME				HARDWARE	**10**	**11**
NO.	SIZE	RATING	TYPE	TYPE	HEAD	JAMB	SILL	GROUPS	REMARKS	REV
										△
										△
										△
										△
										△
										△
										△

Figure 3.35 *Sample Door and Opening Schedule.*

WINDOW, LOUVER, AND PANEL SCHEDULES

Windows, louvers, and building panels are often represented by schedules. The primary reason for scheduling these items over other graphic techniques is to take advantage of their repetitive forms. Assume a building has thirty-eight windows, but there are really only six different configurations of windows. These six windows can be drawn in a schedule where amplifying data is added. Do the same where a building is fabricated from precast concrete panels. The number of "different" panels is much less then the total number of panels. So, as you can see, the primary reason to use a window, louver, or panel schedule is to group repetitive elements in a single location where amplifying information about size, material/finish, and installation can be shown.

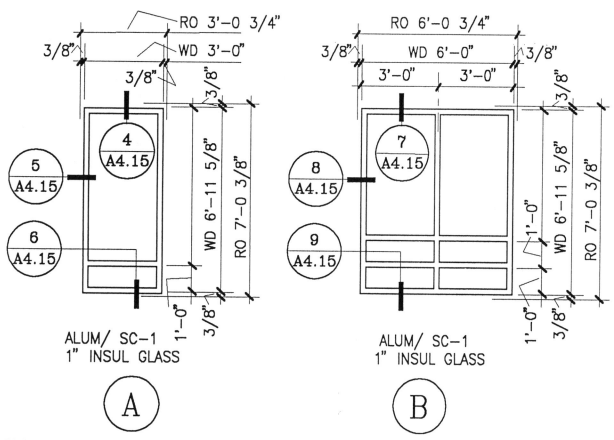

Figure 3.36 *Example Window Schedule.*

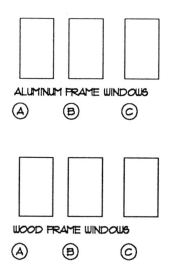

Figure 3.37 Group windows by material and finish.

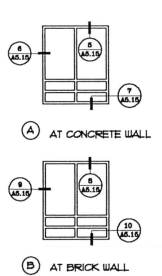

Figure 3.38 Draw new type for each condition of wall material.

Window Schedule

The *Window Schedule* is a familiar tool for expressing window requirements. It begins at the design phase when exterior elevations are being studied. This is the time when basic window shape and overall configuration are considered. Often driven by current trend, a window might be square or rectangular, with horizontal or vertical muntins, or both. Many window designs can be tested on the overall building elevation until the desired "look" is achieved. The resulting windows can then be drawn in schedule form.

Group Windows by Material and Finish

The first step is to group windows by material and finish. For example, a building might have both clear anodized aluminum windows and vinyl-clad wood windows. Keep them separate from each other. This will make it easier to keep track of conditions while doing the design, and will help the contractor find all windows of a kind for estimating and construction.

Create Unique Frame Types for Variations in Wall Condition

Give every condition of frame installation its own window type. You might have two or even three windows that look basically alike, except for the way in which they are fitted into the building. For example, the first window might fit into a concrete wall and will therefore have a frame made to fit into a concrete wall. The second window might fit into a stucco wall with a frame quite different from that used for a concrete wall. In the interest of clarity, give each condition its own window type. Avoid saying, "B same as A except. . . ." The exceptions can often grow out of control.

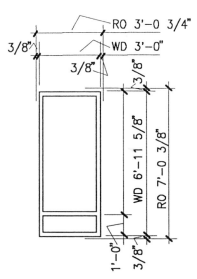

Figure 3.39 *Window dimensioning.*

Reference Details on the Elevations

Once the elevations have been drawn, place detail references at every head, side, and sill condition. Each window type should be unique, so there should be only one set of detail references on each elevation. If conditions seem to require more than one reference along any side, determine whether this is because the rough-opening material changes or because this is really another window type. If the later applies, add another window type to help keep the picture clear.

Dimensioning

Dimension the window, and its rough opening (see Chap. 2, "Dimensioning," for assistance). Since the rough opening for a stucco wall might be different from the rough opening for a concrete wall, a separate frame type is needed to clearly show each condition. Again, avoid saying, "Frame B same as A except. . . ." With the unique frame type, a clear and unique set of dimensions can be shown.

Identify Glass Types

Indicate the glass type for every frame at the frame title when all glass in a given frame is the same. When some glass is tempered and other glass is not, and they are in the same frame, then place a glass-type indicator in every glazed opening in the frame. There must be no doubt about which glass to use in what opening.

Complete the Window Schedule by identifying each frame with a consecutive number, prefaced by the letter "W" for window, all placed in the standard ½" circle. Above each I.D. letter, denote frame material/finish and glass type used in that frame.

Continue this process with every condition in the project. Start with the most typical and proceed to the unique.

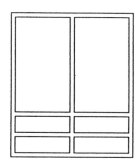

ALUM/ SC-1
1" INSUL GLASS

Figure 3.40 *Note glass type.*

Special Conditions

Some conditions are not generally scheduled because they are unique. A storefront is one example. There might be only one configuration of frame, door, and glazing that looks just like this one. However, it should be treated the same as all other window-like assemblies and shown on the Window Schedule. The same applies for greenhouse rooms and skylights. The idea here is to provide basically two places for the contractor to find every framed and glazed opening on the project: the Door and Opening Schedule and the Window Schedule.

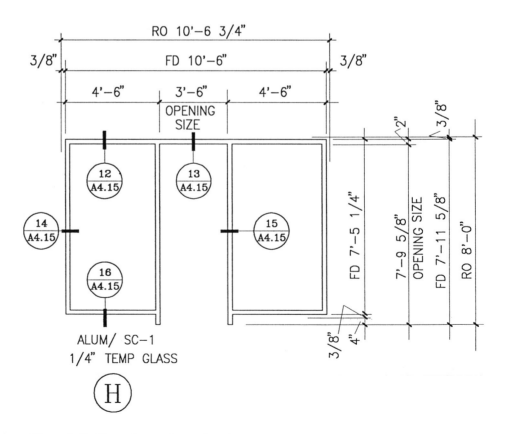

Figure 3.41 *Example storefront elevation.*

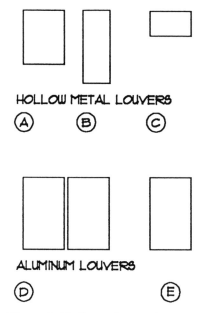

HOLLOW METAL LOUVERS

Ⓐ   Ⓑ   Ⓒ

ALUMINUM LOUVERS

Ⓓ   Ⓔ

Figure 3.42 *Group louvers by material and finish.*

LOUVER SCHEDULE

Another element well suited to scheduling is *louvers.* Louvers are drawn on building elevations, but this is no place to show the fine points of their installation. At ⅛″ scale, it is often difficult to do more than place one reference symbol to call out the louver type. All specific information can more easily and much more clearly be shown on a Louver Schedule.

Group Louvers by Material and Finish

Group all louvers by material and finish. For example, there might be a number of extruded aluminum louvers at the penthouse and some hollow metal louvers in the airshaft. These two materials should be grouped within the schedule to make each louver easier to find.

Reference Details on the Elevations

Place detail references at head, side, and sill conditions, and any intermediate conditions that require field assembly. If a condition seems to require two sets of installation details because of different wall materials, perhaps a new louver type is needed. The frame condition may not be the same for differing wall materials, so don't try to force the schedule to describe those conditions within a single louver type. When two different louvers exist, unless they really are the same, draw a new louver type for each.

Create Unique Louver Types for Variations in Wall Conditions

For like conditions of wall material and louver installation, a number of sizes can be represented on one louver elevation. Each dimension condition must, however, receive its own "type" indicator. If absolute clarity cannot be achieved on one picture, draw individual louver types for each. Dimension the louver and its rough opening. Check for differences in rough opening for each wall condition and draw a new louver type if they are different.

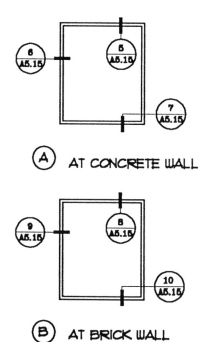

Ⓐ AT CONCRETE WALL

Ⓑ AT BRICK WALL

Figure 3.43 *Prepare a unique louver type for each condition of wall material.*

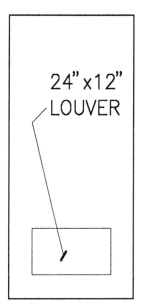

Figure 3.44 *For small louvers in door frames, illustrate in door package.*

Assign Louver Identification

Complete the Louver Schedule by identifying each louver type with a consecutive number prefaced by an L as in L-1, L-2, L-3, etc. Place this number in a standard ½-inch circle located under each louver type. Above the louver indicator, note the material/finish.

Special Conditions

For some conditions, the louver may be a part of another opening such as a window or hollow-metal assembly. When this is the case, include the louver in the appropriate schedule. For example, there could be a single louver located over a door, and all enclosed by the same hollow-metal frame. This condition should be detailed under the door-and-opening package. The frame type would show detail references for all members including the louver. It would also show dimensions and material/finish note.

In the condition where an array of louvers also contains a door, the elevation is mostly louver and should be drawn in the Louver Schedule. The door is still referenced in the Door and Opening Schedule, as are door type and frame type. The frame type, however, becomes the louver type and is noted that way in the frame-type column. (For more information about referencing your way through a set of drawings, see Chap. 9.)

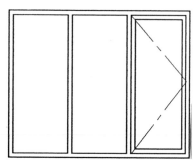

Figure 3.45 *For arrays of louvers, illustrate here, even if they include a door.*

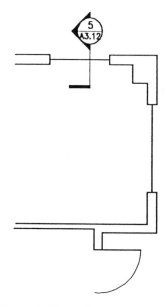

Figure 3.46 Wall section.

PARTITION SCHEDULE

Walls and partitions for a building can have many components and perform a variety of functions. The method for showing these requirements are basically two:

    ###   Draw wall sections.

    ###   Draw a Wall and Partition Schedule.

Wall Sections

Wall sections are usually drawn to describe exterior walls and interior bearing walls. They extend from the foundation to the roof and show all combinations of materials and assemblies along the way. Some wall sections, as through a concrete shear wall in a parking structure, are very simple. Other wall sections, as through the exterior wall of a Victorian house, are very complex. What they all have in common is the fact that the wall section shows how the floor systems and walls come together and, for exterior walls, how the wall passes the floor line.

Partitions

Partitions generally describe interior assemblies that simply divide space and conceal utilities. Partitions often are required to provide fire separation between two spaces as in corridor walls and elevator shafts. Partitions extend from the floor and stop somewhere below, at or above the ceiling. Some partitions extend to the underside of the floor plate above. Partitions are generally of wood or metal stud construction with gypsum board or plaster surface material.

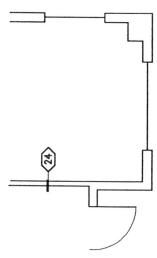

Figure 3.47 Partition type.

Separate Walls From Partitions

Before beginning a Wall and Partition Schedule, determine which conditions will be shown by drawing wall sections, and which will be shown in schedule format. There are two basic elements to consider:

If the wall is structural or extends from the foundation to the roof, draw a wall section. Do this for both interior and exterior walls.

If the wall or partition is of concrete or masonry construction, draw a wall section.

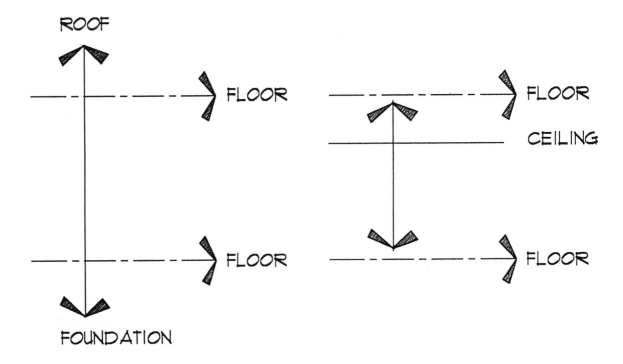

Figure 3.48 Wall sections.   **Figure 3.49** Partitions.

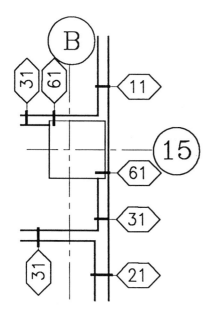

Figure 3.50 *Partial plan showing partition types.*

Partitions by Category

Most other conditions can be shown on the Partition Schedule. The Wall and Partition Schedule is divided into 10 categories. These 10 categories are used to group assemblies of like construction and purpose. These are the following:

01–09	Basic partitions with no fire rating. May have acoustic rating.
10–19	1-hour fire-rated assemblies. May also have acoustic rating.
20–29	2-hour fire-rated assemblies. May also have acoustic rating.
30–39	Shaft assemblies, 1-hour and 2-hour rated. May also have acoustic rating.
40–49	Chase assemblies, nonrated and 1-hour fire-rated assemblies. May also have acoustic rating.
50–59	Additional chase assemblies when needed.
60–69	Furring assemblies. No fire rating. May have acoustic rating.
70–79	Additional furring assemblies.
80–89*	Concrete and masonry assemblies.
90–99**	Job-specific assemblies

*Concrete and masonry partitions may be included when they are not otherwise covered by wall sections or structural drawings.

**The last category is for assemblies which do not fit the other nine categories. This is used for remodel work where assemblies like solid plaster or plaster on wire studs are reconstructed.

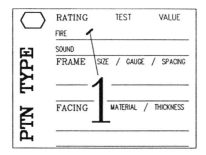

Figure 3.51 *Partial Partition Schedule form.*

Fire Testing Agencies

FM   Factory Mutual Research Company

NBS   National Bureau of Standards

UL   Underwriters Laboratories, Inc

UC   University of California

Figure 3.52 *List of testing laboratories.*

Partition Schedule Format

The Partition Schedule Format has two components. The left side is a fill-in-the-blank form calling for the following:

Fire and acoustic rating

Framing material, size, and spacing

Surface material

1 Rating:
There are two lines for rating information. The first line is for fire rating. Enter the rating, i.e., 1-hour or 2-hour, followed by a test-report indicator which proves the assembly actually meets the rating required. This test report number can be one of several options. First, Table 43 of the Uniform Building Code lists assembly requirements. Simply enter the appropriate number on the form. Fire tests have been conducted by private laboratories including Underwriters Laboratories and Factory Mutual to name only two. Their test number may be used in lieu of the UBC-Table-43 number. Another source, and often the simplest to use, is the Gypsum Association schedule of gypsum board assemblies for walls, furring, and more. For each assembly, a whole list of test reports is given. The only entry needed on the Wall and Partition Schedule is the "WP" number from the Gypsum Association manual.

GYPSUM ASSOCIATION
TEST ASSEMBLY
WP1070

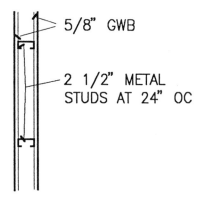

5/8" GWB

2 1/2" METAL
STUDS AT 24" OC

Figure 3.53 *Test assembly from gypsum association.*

Acoustic Rating

The second line under "RATING" is for acoustic requirements. Enter the value required, i.e., 45STC, followed by the Gypsum Association "WP" number for the assembly being used.

A wall or partition which has both fire and acoustic ratings must have the same "WP" number in the "TEST" column of the schedule form. The contractor can not build an assembly with two different requirements.

Variations on Tested Assemblies

There are many cases where the wall or partition assembly needed does not exactly match the assembly shown in the test reports. One such example is stud size. Most agencies will regard as equal a stud size that is greater in width or gauge than the one required by the referenced test report. Because of this, tests are conducted on the minimum value of stud and covering. For example, 3½-inch-by-20 gauge studs at 16-inch centers are common in construction. The test might, however, use 2½-inch-by-25 gauge studs at 24-inch centers. If this passes the test (and it does), the better wall can be used too. Always look for a tested assembly that is the same or less restrictive than the condition you intend to use. There may be exceptions to this general rule, so if there are any doubts, contact the building office for a ruling before making a decision.

MEETS REQUIREMENTS
OF TEST ASSEMBLY
WP1070

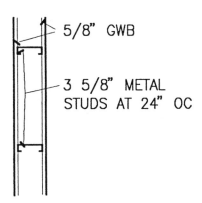

5/8" GWB

3 5/8" METAL
STUDS AT 24" OC

Figure 3.54 *Partition which matches closely enough to test assembly.*

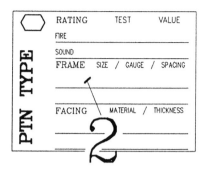

Figure 3.55 *Partial schedule form.*

2 Framing:

The second entry in the Partition Schedule is used to identify the framing member or system. Space is provided for MATERIAL, SIZE/GAUGE, and SPACING. Enter framing material as METAL or WOOD. SIZE is stated as 3½″, 25, or 2 × 6, etc. Note SPACING as 16 or 24. The "inches" is understood.

The next entry is for FACING. Here, MATERIAL and THICKNESS are entered, i.e., GWB and ⅝. Again, "inches" is understood. *Do not* include the finish in this entry. For example, ceramic tile applied thin-set over tile backer board or moisture-resistive gypsum board is *not* included on the partition type. Finishes are described in the Fnish Schedule and not in the Wall and Partition Schedule. To add finish information would increase the potential number of partition types exponentially.

Define Gypsum Wallboard

The term *gypsum wallboard* may be used to represent a variety of products from the standard product used in nonrated construction to exterior sheathing. In the interest of keeping the Wall and Partition Schedule as brief as possible, all the variations of gypsum- and cement-based panels will not be individually included. These will, however, be defined and explained in the general notes and specifications as follows:

   ###   The typical gypsum wallboard is a paper-faced, nonfire-rated gypsum panel of thickness shown on the drawings.

   ###   Where fire-rated gypsum wallboard is required, use Type-X gypsum panel of thickness noted on the drawings.

   ###   When gypsum wallboard is shown in exterior applications, use exterior gypsum board sheathing or soffit board as appropriate in thickness shown on the drawings.

   ###   Where gypsum wallboard is shown in toilet rooms, kitchens, and other such wet areas, use moisture-resistive gypsum panels of thickness shown on the drawings.

3 Graphics:
The right-hand portion of the Wall and Partition Schedule is for drawing a PLAN and SECTION VIEW of the assembly. The plan view should show one framing member and the facing material. Note items that cannot be included in the left-hand form, like acoustic insulation. The section view shows the top of the assembly in very schematic form. Show structure as a single line. The same applies to ceilings. Show the assembly as it is configured at and above the ceiling. On the PLAN VIEW show the nominal thickness of the assembly.

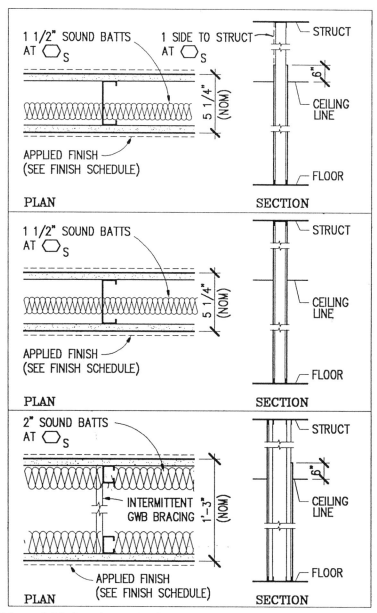

Figure 3.56 *Sample graphic for simple partition types.*

SOME USEFUL TIPS:

The following are tips to consider when preparing a Wall and Partition Schedule.

1 A unique type must be drawn for each of the following variables:

Fire rating

Condition at or above the ceiling

Facing material

2 Unique types are not required to illustrate the following conditions:

Change in stud size or gauge from one already shown. Note the exception by placing the desired stud size next to the indicator symbol on the plan where the exception occurs. If there is a lot of this, draw a new assembly type.

Change in wallboard material is a function of location and finish. A general note is written to explain which material is used in what application. For example, exterior gypsum sheathing is used on exterior surfaces while moisture-resistive gypsum wallboard is used in areas like bathrooms where a lot of moisture is present.

Change in finish. A partition type is not dependent upon the finish applied to it. It can be brick veneer or just paint. The remainder of the assembly must, however, be capable of receiving the finish scheduled. I once tried to use 25-gauge studs to support cement plaster. It didn't work.

Addition of thermal or acoustic insulation. A basic assembly can be used by placing an "A" or "T" next to the indicator symbol on the plan where the exception occurs. This may apply more to acoustic exceptions because thermal requirements are more often shown on wall sections.

Regulate the Use of Exceptions

A partition can quickly become undefined if too many exceptions are piled on top of a basic configuration which was previously defined in the schedule. To avoid this:

Do not make more than two exceptions at any single condition.

Do not use fire-rated partitions when the rating is not required.

Conversely, do not use nonrated partitions where a fire rating is required. Use a new assembly type.

On the graphic portion of the Partition Form there is room to add detail reference symbols to point the reader to the detail for construction. This works for special conditions but is better treated in the Partition General Notes.

CONCLUSION

Complete the Partition Schedule one assembly at a time. Progress through each plan by identifying every partition encountered. As stated in the General Notes, the typical partition types are not referenced on the plans. This saves a lot of time, and leaves crowded drawings open for other information. When the schedule is complete, check to be sure that all plan conditions are properly described.

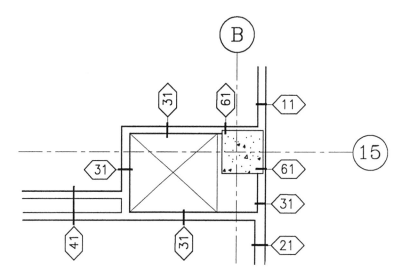

Figure 3.57 *Partial plan showing partition indicators.*

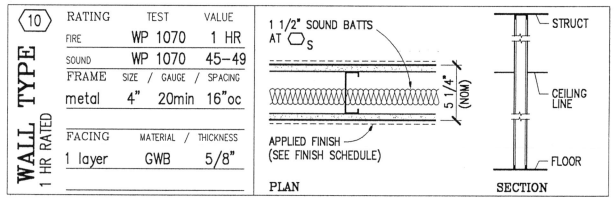

Figure 3.58 *Complete example of a partition type.*

EQUIPMENT SCHEDULE

Most architectural projects have some kind of equipment, either built-in, as in a commercial kitchen; or moveable, as in a computer terminal. The coordination of this equipment so that it fits into the building and has proper rough-in is one of the least-understood of our services to the owner.

Start at Contract Negotiations

The place to start dealing with equipment is at the time of contract negotiations. An agreement must be reached with the owner for the amount of equipment coordination necessary to adequately plan and design the building. The owner may elect to have the architect select and purchase everything from chemical fume hoods to wastebaskets, or to disregard all equipment coordination as unimportant or unaffordable. Usually, some point in between is selected.

OWNER
ARCHITECT
AGREEMENT

Owner Furninshe Equipment

Fixed and Movable Equipment

The Owner shall provide...

The Architect shall ...

Figure 3.59 *Get it in the contract.*

Figure 3.60 *Equipment groups.*

Equipment Groups

To assist the owner and architect in determining scope of service, equipment is divided into three basic groups.

1 Group 1 equipment is also known as "fixed equipment." This equipment is permanently attached to the building and includes items like a walk-in cooler and scullery dishwasher. The architect is responsible for selecting, specifying, detailing installation, and coordinating mechanical and electrical services. Full fee is charged, based on cost of equipment.

2 Group 2 equipment is also known as "major moveable equipment." This equipment group consists of large items having reasonably permanent location in the building, but which are not permanently attached. Examples include furniture, carts, refrigerators, and office machines. Full fee is charged when the architect is responsible for selecting, specifying, detailing installation, and coordinating mechanical and electrical services.

3 Group 3 equipment is also known as "minor moveable equipment." This group consists of small, very moveable items like wastebaskets and staplers. It is rare that much time is spent coordinating items from this group, though it has an impact on the owner's total move-in budget. Because little time is spent, no fee is generally charged for coordination of this group.

The owner may elect to list out her or his own equipment to save architectural fee. This is not recommended. The architect is much better-qualified to survey equipment needs, and will eventually provide the service whether paid for it or not. Owners are specialists in their own field, but generally not in gathering technical data about equipment.

Since some owners insist on providing this information, an *Equipment Data Form* has been designed for their use. This form, accompanied by an instruction sheet, will prompt the user for the correct information. It should be noted here that whether the information is gathered by the owner or the architect, it is very important to be thorough and accurate when filling in the form. Faulty information can cause very expensive problems during construction. The following description will assist everyone involved to properly complete the form.

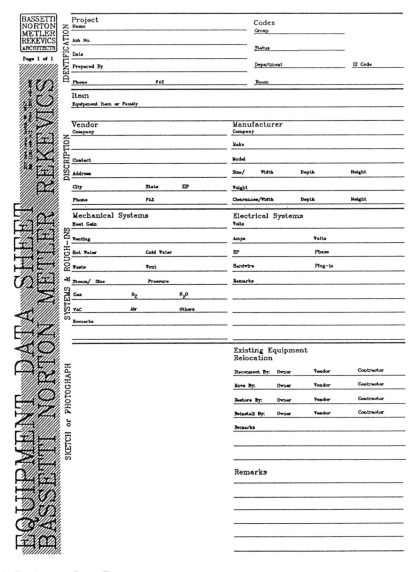

Figure 3.61 Equipment Data Form.

THE EQUIPMENT DATA FORM

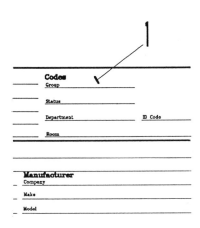

Figure 3.62 *Equipment Data Form.*

Code and Group Status

1 This area is for equipment identification. CODE and GROUP will be assigned by the architect. Enter STATUS from the list below.

F.I.C.   Furnish and Install by Contractor.

F.I.O.   Furnish and Install by Owner. For existing items note F.I.O. (E.) and enter existing location in RE-MARKS area. Note future items F.I.O. (F.).

F.I.T.   Furnish and Install by Tenant. For existing items note F.I.T. (E.) and enter existing location in REMARKS area. Note future items F.I.T. (F.).

F.O.I.C.   Furnish by Owner and Installed by Contractor. Note existing items F.O.I.C. (E.) and enter existing location in REMARKS area.

F.T.I.C   Furnish by Tenant and Install by Contractor. Note existing items F.T.I.C. (E.) and enter existing location in REMARKS area.

F.O.I.V.   Furnish by Owner and Installed by Vendor. Note existing items F.O.I.V. (E.) and enter existing location in REMARKS area. Note future items F.O.I.V. (F.).

F.T.I.V.   Furnish by Tenant and Install by Vendor. Note existing items F.T.I.V. (E.) and enter existing location in REMARKS area. Note future items F.T.I.V. (F.).

Enter the department and room in which this item is to be installed or used. When many rooms will receive the same equipment, enter the room type, i.e., classroom, exam room, or office.

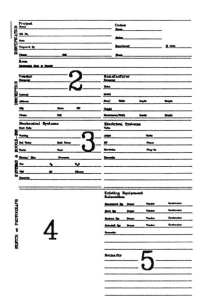

Figure 3.63 *Equipment Data Form.*

2 Name, Make, Model:
Enter equipment-item name such as copier, desk, or walk-in cooler, along with manufacturer, make, and model number. If a vendor is the primary contact, enter vendor's name and address, phone, and fax. If the item comes direct from the manufacturer, enter N/A in vendor space and enter manufacturer's phone and fax. In either case, enter the name of the contact person. Note equipment size by width, depth, and height along with clearance requirements for maintenance and air circulation. Finally, note the weight.

3 Services and Rough-Ins:
Mechanical information needed includes rough-in requirements for hot water (HW), cold water (CW), waste (W), steam (STM), gas (G), and vent (V). Enter each of these as required by individual equipment items. Show rough-in size and fitting requirement. The service will be determined by the mechanical engineer. Some equipment puts out a lot of heat. Enter this heat gain in BTUs.
Electrical information needed includes requirements for voltage (V), amperage (A), wattage (W), horsepower for motors (HP), and phase (P). Note requirements for hardwire (HW) or plug-in (PI), and note any special plugs. Note requirement for dedicated circuit and clean power under OTHERS.

4 Photo or Sketch:
Paste a photograph or sketch or copy from a catalog in the SKETCH or PHOTO space. This is a good place to point out rough-in locations when critical. When no catalog art is available, as for existing equipment, draw a sketch showing critical parameters.

5 Remarks:
REMARKS is for adding information not shown in the blanks above. The information can assist the design team in better planning for installing the item. Some examples of information needed include shielding requirements, noise and vibration, and attachment to building structure.

CONCLUSION

The Equipment Data Form should be accompanied by catalog sheets, manufacturers' brochures, and the like whenever possible. These should be stapled to the back of the form so they don't get lost.

When all needed data is recorded, the architect will assign CODE and GROUP numbers and distribute the information. An Equipment Schedule will be created, and the equipment will be shown on drawings. All future references to equipment should be by equipment number.

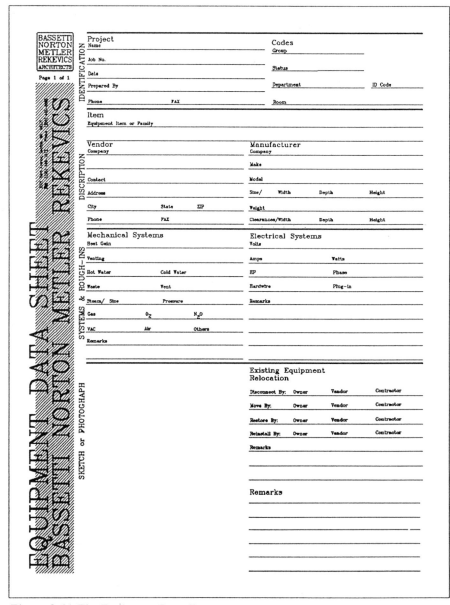

Figure 3.64 The Equipment Data Form.

THE EQUIPMENT SCHEDULE

With the information gathered on the Equipment Data Form, a schedule can be put together listing items which concern the contractor. These include all FOIC, FTIC, FIC, and some FIO, FIT, FTIV, and FOIV equipment which requires support or rough-in. Make entries as they appear on the Equipment Data Form.

1 Column number 1:

Column number 1 of the Equipment Schedule is *the equipment identification code*. This code consists of two letters which represent the equipment category, followed by a consecutive number for each item in a category. Some items are a part of a family of equipment. Data-processing equipment often fits into this category. A mainframe computer will be located in a computer room with remote terminals and printers located where they are needed. In this case, the main item in the family is identified first, with peripherals receiving a consecutive letter suffix.

See figure below for a list of suggested categories followed by some of the kinds of equipment that fits into each.

EQUIPMENT CATEGORY CODES

Ar.	Art	Lb.	Laboratory
Ae.	Athletic	Li.	Library
Ap.	Appliances	Ln.	Laundry
Di.	Diagnostic Imaging	Ma.	Machinery
Dn.	Dental	Me.	Metal Working
Dp.	Data Processing	Pc.	Patient Care
Dt.	Detention	Rf.	Refrigeration
Fs.	Food Service	Wd.	Wood Working

Figure 3.65 *List of equipment categories.*

2 Column number 2, Equipment Status:
This column is for equipment status. Enter FIC, FIO, FIT, FOIC, FTIC, FOIV, or FTIV as noted on the Equipment Data Form. For FIC items, be sure a full description exists in the specifications.

3 Column number 3, Item Description:
This column is for item description. This entry should be brief. Use generic descriptions when the item is to be furnished and installed by the contractor.

4 Column number 4, Mechanical:
This column is for mechanical requirements. List in the same order each time to avoid confusion, starting with hot water, cold water, waste, steam, gas, and ventilation. Be specific and note only rough-in sizes.

ID	EQUIPMENT STATUS	ITEM	SYSTEMS Mechanical	Electrical	REMARKS	REV
Rd.27	FIO	Film Illuminator	N/A	115V	See Detail 27/9.12	△
		Four Panel		15A		△
		Recessed		Hard Wire		△
						△
						△
						△
						△
						△
						△

Figure 3.66 Equipment Schedule.

5 Column number 5, Electrical:
This column is for electrical requirements. Again, list them in the same order each time, starting with voltage, amperage, wattage, horsepower, phase, hardwire, and plug-in. Be specific. Don't assume 408 power. It might be 440, which has different electrical requirements.

6 Column number 6, Remarks:
This column is for remarks. The first entry should always be the reference to installation details. Choose remarks that assist the contractor in supplying and/or installing the proper items.

7 Column number 7, Revisions:
This column is for the revision indicator. When an entry is revised during construction, place the appropriate revision indicator in this column. (See Chap. 8 for more information about revisions.)

EQUIPMENT			SYSTEMS			
ID	STATUS	ITEM	Mechanical	Electrical	REMARKS	REV
Rd.27	FIO	Film Illuminator	N/A	115V	See Detail 27/9.12	△
		Four Panel		15A		△
		Recessed		Hard Wire		△
						△
						△
						△
						△
						△
						△

Figure 3.67 Equipment Schedule.

Complete the Equipment Schedule by listing all FIC, FTIC, and FOIC equipment and the FIO, FIT, FTIV, and FOIV equipment which requires work by the contractor. Prepare the schedule using equipment codes in alphanumeric order.

A schedule such as this works best when produced on a computer. See your local data-base software vendor about the right program for your needs.

EQUIPMENT			SYSTEMS			
ID	STATUS	ITEM	Mechanical	Electrical	REMARKS	REV
Rd.27	FIO	Film Illuminator	N/A	115V	See Detail 27/9.12	△
		Four Panel		15A		△
		Recessed		Hard Wire		△
						△
						△
						△
						△
						△
						△
						△
						△
						△
						△
						△
						△
						△
						△
						△
						△
						△
						△
						△
						△
						△
						△
						△
						△
						△
						△
						△
						△

EQUIPMENT SCHEDULE

Figure 3.68 The Equipment Schedule.

CONCLUSION

The schedules illustrated in this chapter are the ones most commonly used in architectural document production. Many more are used to describe incidental requirements. One such example is the *Loose Lintel Schedule* used to define steel shapes over masonry openings. Schedules like this can be invented any time an array of information is required.

Remember the two basic forms used for producing schedules: *graphic presentation* and *matrix.* Use these two formats any time they can help you present information more clearly and in less space.

GRAPHIC MEDIA

4

INTRODUCTION

The next three chapters deal with the way drawings are produced: the materials used for drawing, the reproduction processes, and drafting systems. This information is aimed at helping you to make the process of producing drawings more efficient and more accurate.

This chapter deals with the products used to prepare today's architectural drawings, from polyester drawing film to mechanical pencils. It describes the most commonly used media so that you can be better-prepared to discuss this subject with your local reprographics shop, and hopefully, to select the right material for the job. Product descriptions are followed by some of the products' limitations which you should discuss with your reprographics manager to determine their impact on your work.

Much of the information in this chapter came to me from Olympic Reprographics of Seattle, Washington, specifically in the persons of Mike Murdock and Sam Mellison. Mike has been selling me architectural products for almost 20 years, and I am indebted to him for his help here and for those many years. I suggest that every architectural office designate one architect to receive vendors like Mike, and maintain a library of current graphics products. Your reprographics manager is an independent businessperson, not tied to any one product line. You are, therefore, likely to get all the information about a product, both good and bad. If one product doesn't meet your needs, there will probably be another product that will. Manufacturers' representatives, on the other hand, are not always as willing to discuss the limitations of their products. A continued relationship with your reprographics manager can help you select the right materials for your practice and solve problems like pen cleaning and plotter skip. It will also, probably most importantly, allow you to stay on top of the latest products and techniques.

This chapter deals with the tools used to produce drawings. They include the following:

sketch paper
drafting and tracing paper
polyester drafting film
plotter media
drawing lead and lead holders
mechanical pencils
technical pens and ink
plotter pens
erasers
appliqués

This chapter will give you a general overview of these products and how they are used.

SKETCHING PAPER

Study sketches are often drawn very quickly, only to be thrown away. They are fast, usually freehand drawings and generally short-lived. They also consume a great deal of paper. Sketch paper is an inexpensive medium which reproduces (with care) in the diazo machine. It is not designed to be used for final drawings where permanence and rapid reproduction are important.

The main attraction of sketch paper is its price. It is only about $1/10$ the cost of polyester drawing film. This makes it essentially disposable and therefore very usable. Sketch paper is an excellent medium for pencil drawings and some felt pens. Pencil lead is easily erasable, even on paper as light as sketch paper, making it possible to change or test several ideas. Using felt-tip pens on this type of paper makes drawings of high contrast. Pencils and felt-tip pens can be used together to indicate shadows and glazing on building elevations. Sketch paper can also be used with other drawing media including chalk and charcoal. The fact that this paper is so very inexpensive and so very versatile makes it indispensable to every architectural office.

Limitations

This product will not tolerate continual reworking and erasures without wearing through. Many felt-tip pens will bleed on sketch paper, so test them before they are used on a good drawing. Sketch paper often gets caught in a diazo printer. If it does, the drawing is usually destroyed. Consider using a flat-bed or vacuum frame (see description in Chap. 5) for copying drawings done on this type of paper.

Stock

Sketch paper is a very lightweight paper. Generally it is in weights of seven pound or nine pound. It comes in both white and canary, with the white being slightly more expensive than the canary. It comes in the following roll sizes:

WIDTH	LENGTH
12 inches	50 yards, 20 yards
14 inches	50 yards
18 inches	50 yards, 20 yards
21 inches	50 yards
24 inches	50 yards, 20 yards
36 inches	50 yards
42 inches	50 yards

Drafting and Tracing Paper

Paper is used as a relatively inexpensive drafting medium for permanent drawings. It is most often used for small projects where extensive rework of the drawing is not expected. It is also used for CAD progress plots which are generally short-lived.

The paper stock used for today's drawing is specially formulated for optimum performance in drafting and reproduction. They are 100 percent rag for drafting vellum, and lesser rag content for tracing paper. Drawing papers have the following qualities:

They are made especially for graphite and clay-based leads and will hold a strong line with a minimum of lead on the paper. This helps keep drawings clean longer.

The 100 percent rag vellums are more durable than sulfite-based tracing paper. They can withstand repeated erasures in the same place without leaving a ghost of the previous work and maintaining a draftable surface ready for more work.

They have permanence. They are nonyellowing and nonhardening.

Some are suitable for pen plotters using liquid-ink-disposable, roller-ball or felt-tip pens.

Limitations

These papers are not well-suited to ink drawing if changes are anticipated. Erasing ink line work often "shreds" the drafting surface. Redrafting in ink over these areas results in feathering. This in turn may not erase at all. On some drawing papers the only way to remove an ink line is to cut it out. Further, many felt pens and markers bleed through or feather on this material, even on sound surfaces. If you want to use ink, test the products together before doing a whole drawing.

Paper is not dimensional-stable. A drawing taped-down and left overnight may expand enough that traced line work will no longer align perfectly. This makes paper medium a poor choice for overlay drafting.

Stock

Drafting papers come in a variety of weights. Following is a description of two of the most common weights.

> ###   Tracing paper is used for an inexpensive medium where permanence and durability are not required. Consider this stock for drawings which need some amount of erasure and yet will not be used for a long period of time.

> ###   16-pound or 20-pound vellum is used for final drawings which are destined to see many reprographic cycles and eventually the archive. It is generally a hard-finished rag paper.

Each of these papers can be printed with grids in a fade-out blue color. The following are common grid sizes. Many more grids and line patterns are available for specialty applications.

> ### 10×10 squares-per-inch for engineering drawings and site work.

> ### 8×8 squares-per-inch for architectural drawings. This grid can be used as a scale when preparing sketches, and acts as a guideline for lettering.

> ### 4×4 squares-per-inch for detail work at $3'' = 1'\text{-}0''$ scale, and for plans at $\frac{1}{4}'' = 1'0''$ scale.

These papers can be purchased in precut sheets in standard and popular sizes. They are also available in roll stock in the following sizes:

WIDTH	LENGTH
24 inches	20 yards
24 inches	50 yards
30 inches	20 yards
30 inches	50 yards
36 inches	20 yards
36 inches	50 yards
42 inches	20 yards
42 inches	50 yards

Some of the smaller precut sheets are also available in pad form.

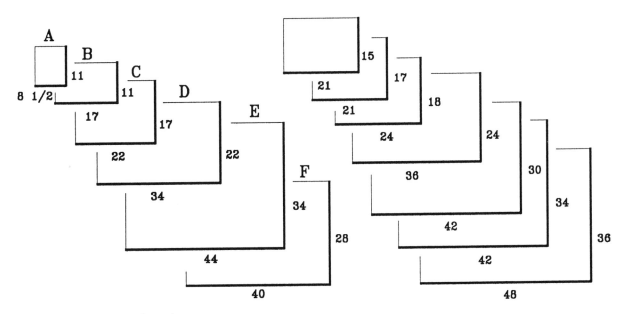

Figure 4.01 *Drawing-sheet sizes.*

POLYESTER DRAFTING FILM

Polyester drafting film has become the standard drafting medium for architects and engineers. It is used for drawings considered permanent, such as those presented at the conclusion of schematics, design development, and contract documents. It is also used where the length of time spent on a sheet might easily wear out a lesser product. It is a stable medium for overlay drafting and automated pen plotters.

There are many varieties of polyester drafting film. By selectively shopping, you can choose the combination of qualities needed for your work. Some things to look for include the following:

Dimensional stability. The drawing stock will not "move around" with temperature and moisture changes. This makes polyester drafting film suitable for overlay drafting.

Fadeproof. The material will not yellow or loose its transparency.

Will not become brittle. Drawings which are rolled-up for years can be unrolled, often laying flat. They will not crack with time.

Tough. Polyester film will not tear or rip. Drawing sheets can take substantial abuse without damage. Drawings last through projects that take over a year to produce.

Static free. Drawings won't stick to each other or to the cylinder of a diazo printer. Dirt and erasure crumbs don't adhere to the sheet so that drawings stay clean longer.

Durable drafting surface. The matte finish is designed to withstand repeated erasures when performed correctly. With proper erasure, the drawing surface should last the life of the project.

               Note: Use either a moistened vinyl or a chemical-imbibed eraser for removal of ink lines and a dry-vinyl for the erasure of plastic lead. Never use plastic plugs in an electric eraser on drafting film without wetting the eraser first.

High translucency. This attribute is important when used for overlay drafting. The higher-translucency film permits printing through several layers at once with increased clarity. For optimum results, do not attempt to print with more than three total layers of drawing film.

Drafting surface accepts lines of combination lead, plastic lead, and ink. The results obtained from using each of these is markedly different and is discussed later. The matte finish is not suitable for work with most felt-tip pens and ordinary ball-point pens. However, some manufacturers have developed a product that works equally well for manual drafting and for automated pen plotters. This may reduce the amount of drawing stock you need to keep on the premises.

Clear drafting films are available for overlay work. They come with a chemical matte draftable surface.

Limitations

Although polyester drawing film is available in thicknesses as thin as 2 mil, this is not recommended for overlay work. The material is too thin for the dimensional stability required by overlay drafting, and the holes can be ripped out when the sheet is pulled off the pin bar. Overlay film should be 4-mil for longer life greater dimensional stability. 4-mil film is also more durable so, the pin-registration holes will keep their size and shape longer.

The grid on pregridded stock is not registered one sheet to another. It is, therefore, not a good choice for overlay drafting. It may be used for base sheets, but remember that each sheet grid is not in the same relative position on the sheet. A second floor drawn over the preprinted grid will not align with the first floor drawn on its own sheet grid. I consider this as a major problem, and do not recommend pregridded drafting film for sheet stock. There is one exception: The detail sheet with a registered 8 × 8 grid is handy for detail work. The key here is to have a print shop reproduce an 8 × 8 grid onto previously ungridded drawing stock. Pregridded stock is still not registered.

Be sure to use antistatic drafting film. Normal drawing film sticks to everything. It is impossible to remove or to replace single drawings in a flat file and originals will likely be eaten by your diazo printer.

With polyester drafting film, overerasing is difficult but not impossible. A worn-out drafting surface will not accept pencil of any kind and only some ink products. Be very careful with electric erasers.

The matte drawing surface on film has a whitening agent added to the chemical matte finish to achieve visual contrast while drafting. The whitener is not necessary for the surface to receive ink or pencil. Film is available in clear matte, but I don't recommend using it. It is very much like drawing on glass. The surface you see is the table top, not the film. This could invite many drafting problems.

When double-matte drafting film is used, a vertical hanging file is a must. Drawings stored in a flat drawer will "grind" on each other as they are filed and retrieved. This causes the drawn line and lettering to wear off. Double-matte drafting film also restricts the use of appliqués. With the added matte thickness, they are forced away from the front-line work causing cloudy reproductions. Appliqués are very difficult to remove from double-matte film, and cleaning residue might remove the matte finish.

Stock

The gauge or thickness of polyester drafting film is measured in millimeters. They include 2, 3, 4, 5, and 7 mil. The common material used for most permanent drawings is 3-mil and 4-mil film. 4-mil film is used in overlay drafting to maintain dimensional stability and to prevent distortion of the film when it is removed from the pin bar. These sizes offer excellent reproduction because the line work is still relatively close to the reproduction surface. The thicker the film, the fuzzier the line quality. Of the remaining thicknesses, 2 mil is an extremely light material and is generally used only where economy and short life are at issue. 5-mil and 7-mil film are very heavy, but do not reproduce well. They are not generally used, and are difficult to obtain without special order.

Polyester drafting film can be ordered with a grid of fade-out blue in the following grid sizes:

10×10 squares-per-inch for engineering drawings and site work.

8×8 squares-per-inch for architectural drawings.

This grid can be used as a scale (loosely), and works as a guide for lettering.

Precut sheets are available in standard and popular sizes. They are also available in a limited number of roll stocks depending on the manufacturer. These include the following sizes:

Standard Sheet Sizes

8 1/2 x 11
11 x 17
15 x 21
17 x 21
24 x 36
30 x 42
34 x 42

Figure 4.02 Standard sheet sizes.

WIDTH	LENGTH
30 inches	20 yards
36 inches	20 yards
42 inches	20 yards

Tip: Gridded film on rolls is not easy to find and is rather expensive. But, if your reprographics supply house will sell you 6-feet-x-36"-wide, place it on your drawing board as a cover sheet. Then use ungridded film to draw on. The background grid reads through so that it can used as a rough scale and for lettering guides. By taping a pin bar to the surface, all drawings register to the grid. *Caution:* Be sure to place the grid side down in case you spill eradicating fluid on it. This will likely remove a portion of the grid.

Polyester drafting film comes with matte surface on one or both surfaces. Single-matte film is for drafting one surface only. Double-matte allows drafting on the back side as well. This is for information-like grid lines which you do not want to erase while making changes to the front. Single-matte allows for the use of appliqués which are always applied to the back surface. This is not as successfully done on double-matte mylar.

Polyester drafting film can be ordered with a row of holes along the top edge for pin-bar registration. These prepunched sheets are the most convenient way to register drawings, as they are produced for overlay drafting. Drawings can be postpunched, but this process allows room for error in aligning the work. If you know in advance that the drawing will be produced with overlays, have prepunched stock at hand.

Any of the polyester drafting films can be ordered with a preprinted title block and border. Or, if you wish, design your own title block and furnish camera-ready artwork to your reprographics house for printing.

Figure 4.03 *Prepunched film.*

PLOTTER MEDIA

Pen plotters and computer-aided drafting and design have had their influence on the paper and drafting-film industry. Products have had to be developed specifically for the high speed of the plotters, and to accept a wide variety of pen types.

Plotter papers come in both bond and vellum. Bond is 16 to 24 pounds in weight. It is available in opaque or translucent types which are used for quick plots and progress drawings where durability is not an issue. The translucent bond can also be diazo-copied. Vellum is 15 to 20 pounds in weight. This product is used when originals must be more durable. When colored pens are used, a presentation-quality drawing can be produced which is also reproducible.

Plotter film is usually double-matte. The reason is that single-matte film does not drive evenly through the plotter. The wheels that propel the film often slip on smooth film's back surface, causing the drawing to be distorted. The back surface on plotter film stops this slipping.

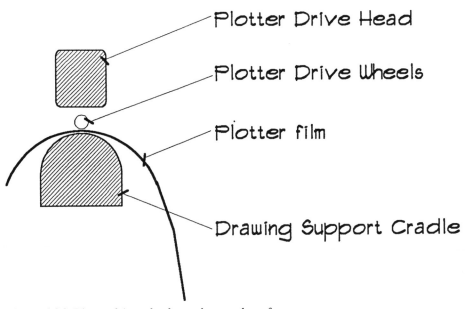

Figure 4.04 Plotter drive wheels need a rough surface.

Limitations

As with all paper products, especially where plasticizers are used, they are not considered permanent. Lightweight papers are not very durable. If a lot of prints are needed, the drawing might be damaged. Drawings stored for awhile in roll tubes will be difficult to unroll.

Stock

Plotter paper and film come in a wide variety of sizes and finishes. The stock most commonly used is usually readily available.

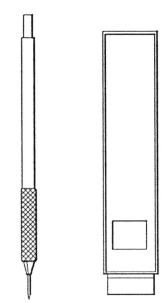

Figure 4.05 Lead holders.

DRAWING LEAD AND LEAD HOLDERS

Drafting lead and lead holders are designed for the production of architectural and engineering drawings. They are manufactured in a large variety of styles of lead holders and qualities for lead, to give users a good selection from which to choose their favorites.

Lead Holders

There are two basic types of lead holders. The older types, known as *lead holders,* have the chuck which holds the lead on the outside. The newer types, sometimes called *mechanical pencils,* have the chuck inside the holder.

Lead holders use a 2-mm-size lead which works well for drawing freehand and for drafting. However, the lead, because of its thickness, needs constant sharpening to maintain a workable point. Lead holders have a push button at the top end which extends the chuck, releasing the lead. Lead can then be changed, repositioned, or retracted into the barrel of the lead holder so it doesn't mark while not in use. The lead holder is made of aluminum, plastic, or steel. Rubberized or knurled grips work best and offer greater line control. For good line control, the center of gravity should be near the tip. Not all lead holders provide this quality. Lead holders have few design problems, and therefore, almost never break down. I have seen lead holders in constant use for over 25 years.

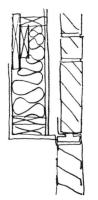

Figure 4.06 Drawing with hard lead.

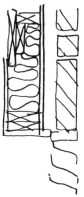

Figure 4.07 Drawing with medium lead.

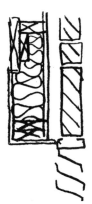

Figure 4.08 Drawing with soft lead.

Lead

The lead used in lead holders come in many varieties depending on the intended use. Graphite and clay leads are used for drawing on paper and vellum. For drafting film, leads are generally polymer-based and plastic-based polymer. These leads provide a dense black line. When controlled properly, they can look almost as good as ink lines.

Graphite leads are graded as hard, medium, or soft and further marked as follows:

Hard lead	9H, 8H, 7H, 6H, 5H, 4H
Medium lead	3H, 2H, H, F, HB, B
Soft lead	2B, 3B, 4B, SD, 6B, 7B

Hard leads, sometimes referred to as "finishing nails" because of their hardness, offer extreme accuracy in drafting. They are used for guide lines, construction lines, dimension lines, and note leaders. Because of their hardness, they can cut the drawing paper like a razor blade, so care must be used. They are also very light, and consequently do not reproduce well.

Medium leads offer the best qualities for drafting on vellum. The harder leads in this group are about at the limit of being reproducible. The softer leads tend to smudge, making the drawing dirty. The most common leads for vellum drafting are 2H for fine lines, H for lines in section, and F for lettering.

Soft leads are seldom used for drafting; however, they are excellent for sketching. Because of their softness, they work well to study building exterior massing and fenestration.

Plastic-based leads have similar properties of hardness to graphite but are often identified by a number system unique to the manufacturer. When purchasing these leads, test them first, if possible, to determine for yourself how they rank.

Drawing leads also come in different colors, including blue, green, red, nonphoto blue, purple, sky blue, vermillion, white, and yellow. Color leads can be used in place of color pencils for marking corrections on prints. Nonphoto blue is used for base-line work and quick notes that you don't want to reproduce.

MECHANICAL PENCILS

Mechanical pencils were introduced in the 1950s and were used by the general public for years before they caught on in architecture. With the development of fine leads (as small as 0.2 mm) and improved plastic leads, the mechanical pencil has now, in many cases, replaced the lead holder.

Mechanical pencils have an internal chuck which advances the lead through an extending sleeve. The sleeve holds the lead so that it does not need to be advanced very far during use. This allowed for the development of thin leads which do not require sharpening. Inside a mechanical pencil is space to store about a dozen spare leads which advance, one at a time, as they are used up. Some mechanical pencils have a cushioned spring tip which helps to eliminate lead breakage.

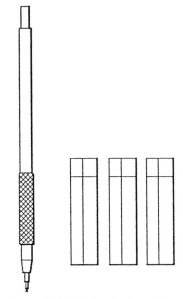

Figure 4.09 *Mechanical pencils.*

Leads

Leads made for mechanical pencils are specially formulated for their thin size. These sizes include 0.2 mm, 0.3 mm, 0.4 mm, 0.5 mm, 0.7 mm, and 0.9 mm. The lead sizes most commonly used are 0.5 mm and 0.7 mm.

Graphite leads are available from 5H through B, depending on lead thickness. Leads for polyester film come in only three hardnesses, and are labeled according to the manufacturer. Color leads are also available including yellow, red, blue, green, brown, and nonphoto blue.

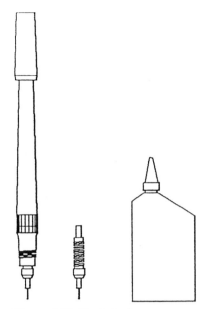

Figure 4.10 *Technical pens.*

TECHNICAL PENS AND INK

Technical pens and drawing ink have evolved to the point to where drafting with ink can now be more precise than ever before. Pens are designed to produce a perfect line every time through a three-part tip. The tip is made with a fine metal filament which controls ink flow. This filament is inside an outer shell which determines the line width. The outer shell is tipped with a wearing surface. This wearing surface is designed for different drawing media. Stainless steel is used on tracing paper or vellum, which has a relatively smooth surface. Tungsten carbide is used on polyester drafting film, which has a much harder surface than paper and actually acts like sandpaper on pen points. Jewel points are also designed for use on polyester film. They are the longest-lasting points, up to 30 times that of stainless-steel points, and provide a smoother drawing point.

Pens come in 12 basic sizes with some amount of disagreement between manufacturers as to what the point size actually is. Notice that point sizes 1 and 2 vary, making the point from one incompatible with the point of another of the same number. Also notice that metric point sizes are in the geometric progression of the square root of 2 (1.414).

Line quality from a technical pen depends a great deal on the ink used in the pen. This is an area where much research is now being conducted to develop inks that flow smoothly, dry quickly on the drawing, yet do not dry in the pen point. Look for ink formulated for technical pens to be used on polyester drafting film.

METRIC SIZE	MARS	CASTELL	KOR I NOOR
.13mm	5x0	4x0	6x0
.18mm	4X0	000	WX0
.25mm	3X0	00	3X0
.30mm	00		00
.35mm	0	0	0
.40mm		1	
.45mm	1		
.50mm	2	2	1
.60mm			2
.70mm	2–1/2	2.5	2–1/2
.80mm	3	3	3
1.0mm	3–1/2	4	3–1/2
1.2mm	4	5	4
1.4mm	5		6
2.0mm	6		7

Figure 4.11 *Pen-point comparison chart.*

Limitations

Drawing pens have one main problem: They dry up. The ink dries up in the point, rendering the pen inoperable. Proper care and cleaning can minimize this problem, but for most of us, pen-and-ink drawing has that one nagging problem. Ink manufacturers are working on formulas to minimize pen clogging. Pen manufacturers are also working on designs that utilize ink cartridges which are prefilled. These cartridges are pressure-balanced with the point to minimize ink "blobbing." Another problem with pens is the fact that they are hard and often scrape the matte finish off the polyester film. When this happens, the material enters the pen point causing the tip to bleed.

Many pen-drafting problems can be prevented through proper maintenance. Here are some tips for storing and using pens:

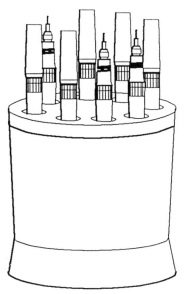

Figure 4.12 *Store pens tip up.*

Store pens with the tips up. Some manufacturers even make their pen holders in such a way that this is automatic.

Always replace pen caps when the pen is not in use. Modern inks will afford a longer cap-off period, but it is still better to replace the cap, so you don't get stuck during a busy period with pens that have dried up.

Purchase good pens of the latest technology. They are getting better all the time.

When pens do dry up or bleed through, cleaning will usually fix the problem. Follow instructions packaged with the pen or try the procedure described below. Pens are extremely fragile, and great care must be used or the pen will be destroyed.

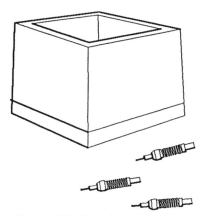

Figure 4.13 Pen cleaner.

Suspend the entire point, while it is still installed in the pen, into a solution of detergent and water, or a solvent recommended by the manufacturer. Allow it to soak for about 24 hours.

Remove the pen point from the pen holder by attaching the point wrench and turning counter-clockwise.

Disassemble the point, which usually consists of an outer case and point, an inner plunger, and an end cap.

Clean all parts.

Insert the plunger into the outer case and run it back and forth a few times to loosen dried ink from the internal surfaces. Be very careful not to bend the plunger. If it is bent, you might consider purchasing a new point.

Rinse the outer case, inside and out. Fill the case with warm water, and run a cotton swab into the wide end, forcing water out of the point.

Dry all parts, being careful not to leave traces of cloth fibers on the pen parts.

Reassemble the point.

PEN PLOTTERS

Plotters used in computer-aided design have unique requirements for pen and ink. First, since the "hand" that holds the pen does not have five fingers, the pen does not need the same shape as pens used for manual drafting. Many pens for plotters resemble mere points which snap into a carrier which is integral to the plotter.

There are three basic kinds of plotter pens. These are:

Ball-tip pens
Fiber-tipped pens
Liquid-ink pens

Of these, the ball-tip and pressure-ball-tip are for check plots where line quality is not important. Plotting speed is very fast with these pens, producing a quick drawing for review of content, but not presentation. Pens are available in one point size, 0.3mm and four colors: red, blue, green, and black. Ball pens are only for use on plotter paper.

Fiber-tipped pens are another inexpensive, disposable plotter pen. These pens are also used for check prints on plotter paper, but are even more versatile. They come with either water-soluble, non-permanent ink for check prints on plotter paper, or solvent-based permanent ink for glossy material like overhead-projector film. These pens are available in two point sizes: 0.3mm and 0.7mm. There are up to eight colors, including black, brown, green, blue, violet, red, orange, and yellow.

Liquid-ink pens are the equivalent of manual technical pens for quality and variety of line work. Points come in 12 sizes from 4×0 to 4. Points are available in four varieties. These are stainless steel, for paper plots; tungsten carbide, for paper, vellum, and coated films; high-speed tungsten carbide, which has cross grooves to allow greater ink flow and faster plotter speeds; and jewel-tipped, for finished drawings.

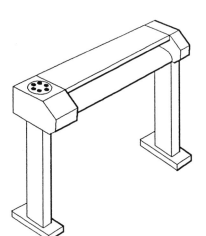

Figure 4.14 Pen plotter.

Disposable pens are replacing the standard liquid-ink plotter pens. They eliminate the mess while providing most of the amenities. Points are available in four sizes: 0.25, 0.35, 0.50, and 0.70. Colors are also available including black, blue, green, and red. Pens are recommended for use on bond, chart paper, copier stock, vellum, natural tracing paper, and drafting film.

Limitation

Plotters can only draw as fast as their pens will allow them to. 20-inches-per-second to 40-inches-per-second pens are considered to be fast pens, while 6-inches-per-second to 12-inches-per-second pens are slow. This is a factor to consider when scheduling CAD plots. They usually take a lot longer than you want them to.

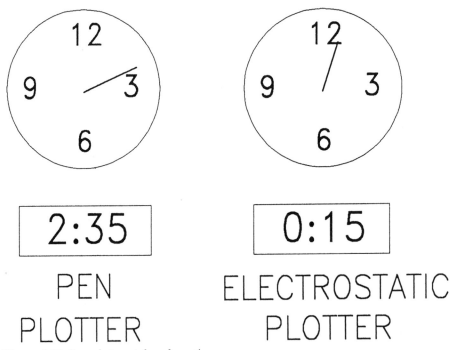

Figure 4.15 *Pen plotters take a long time.*

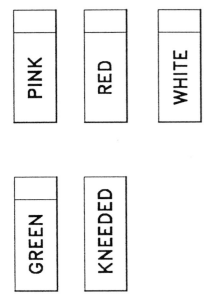

Figure 4.16 *Erasers come in many varieties.*

ERASERS

Erasers are made to remove pencil or ink work from a drawing surface. Since there are so many kinds of lead and ink, and just as many kinds of drawing surfaces, there is more than one kind of eraser. Each has been designed for a specific erasing problem, and should not be used where it is not intended.

Soft pink rubberized erasers or their darker red cousions are made for erasing graphite-lead pencil from paper or vellum. Do not use these on drafting film.

White vinyl erasers are used wet to remove ink from polyester drawing film. Do not use these erasers while they are dry on drawing film because it will damage the matte surface if you rub hard enough. Wet erasing removes most drafting leads and ink with minimal rubbing. White vinyl erasers also work very well (dry) on paper, and are suited to sketch paper which cannot take repeated erasures from a pink rubber eraser.

Chemical-imbibed erasers are used for removing ink from vellum and polyester film. When used on paper, they leave the surface ready for continued drafting. When used on mylar, a soft white vinyl eraser should also be used to clean the chemical from the matte drafting surface. Solvent left on the drawing will cause new ink lines to bleed, and it will also clog technical pen points.

There are combo erasers which combine the chemical-imbibed eraser with a soft vinyl eraser. This way, you can erase the ink line with one end and clean the drawing surface with the other end of the same eraser.

APPLIQUÉS

There are times when hand drafting just doesn't show a desired result, is not fast enough to get the job done, or it just isn't clear enough. Sometimes a mechanical product is needed to enhance the drawing, or just plain get it done on time. There is a line of products used for these situations. They are called *appliqués.*

Appliqués are any graphic product which is applied to a drawing by means of adhesive. Many premanufactured products are available. They generally fit into one of the following categories:

> Dry-transfer products
> Tapes
> Screens, patterns, and color sheets

Appliqués can also be made in the office, and fall into a general category referred to as *sticky backs.*

TEXTURE AND SHADING FILM

Of all of the appliqués, *textural and shading film* may have the largest impact on how a finished drawing will look. Textural film comes in a great variety of textures including diagonal lines, cross lines, straight lines, various dot patterns, and even some material shapes like brick patterns and landscaping symbols like bushes and trees. Shading film comes in a dot pattern, and is measured by the number of dots-per-square-inch and the percent of shading. A shading film marked 550-20 has 550 dots-per-square-inch and is 20 percent opaque. This product, by the way, is the one most often-used on reproducible drawings. It can be drawn-over and lettered-over and still be easily read. Larger dot patterns make line work and lettering difficult to see, as do darker patterns. The 550-20 pattern will also retain its readability when half-sized. Patterns with more dots tend to bleed together. This makes them either solid black or lost altogether when reduced in size.

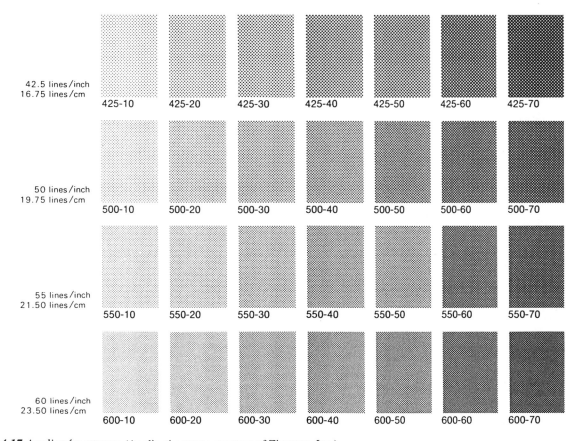

42.5 lines/inch
16.75 lines/cm
425-10   425-20   425-30   425-40   425-50   425-60   425-70

50 lines/inch
19.75 lines/cm
500-10   500-20   500-30   500-40   500-50   500-60   500-70

55 lines/inch
21.50 lines/cm
550-10   550-20   550-30   550-40   550-50   550-60   550-70

60 lines/inch
23.50 lines/cm
600-10   600-20   600-30   600-40   600-50   600-60   600-70

Figure 4.17 *Appliqué patterns.* (Appliqué patterns courtesy of Zipatone, Inc.)

Texture and shading film can be used to indicate many things. They often cover a large area of drawing and are, therefore, placed on the back surface of the drafting sheet. This also allows for continued drafting and changing on the front surface without interference by the appliqué. When preparing an overlay sheet, just for appliqués, they should be placed on the front surface. This will give you better intermediate reproductions. Do not do this if any hand drawing is also required.

Solid-color sheets are also available; these are used to color drawings and posters for presentation, but never on reproducible drawings. Solid colors will not give a consistent gray tone on any reprographic process.

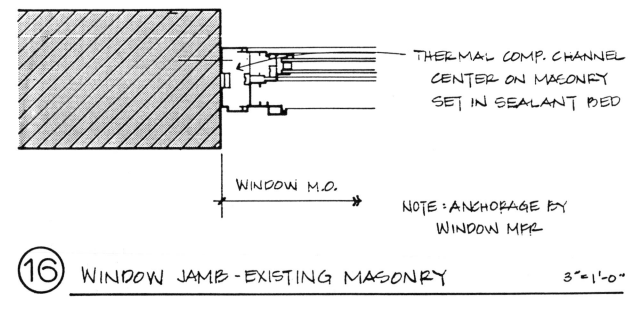

Figure 4.18 *Plan view with 20 percent dot pattern.*

Limitations

Choose patterns that will read the way you want them to. The line work on a diagonal hatch pattern closely resembles a "00" (0.3mm) pen line. If all of your line work is fine, the hatching pattern will dominate the finished product. It will also be very difficult to letter over these hatch patterns. You may need to cut out the patch around notes and titles so that they can be read. This practice can be dangerous if it looks at all like the material represented by the hatch is intended to be omitted in areas where you cut it out. So, select a texture or shade that will yield the desired end result. If necessary, try a few on a test drawing and run some copies. If you have difficulty reading the copies, don't use it. After all, the whole reason for using appliqués is to make the drawing more readable.

Sheet appliqués can be difficult to place on the drawing sheet. It takes skill and practice. If you are doing a job for the first time with appliqués, buy enough extra so that you don't get caught short at 9:30 some night when your office-supply store is closed.

If texture or shading film must be removed, it will take some effort, especially once it has been through the diazo machine a few times. Use the blunt corner of a very fine erasing shield to scrape off areas of film that become difficult to remove. The erasing-shield edges can become razor-sharp after a few years of use and work very well for this purpose. Once the film is removed, clean the drawing-film surface with solvent board cleaner.

Don't use texture or shading film on paper or vellum drawings. Film is more dimensional-stable than paper, so the two fight each other, causing the drawing to pucker or warp, especially after being rolled for awhile. The paper drawings are not as durable as polyester film drawings, and the process of applying these films can damage the paper drawing. The reason is that there is usually a great deal of static caused by pealing off the backer sheet prior to placing the adhesive film. This can cause the drawing sheet, paper or film, to "jump up" and stick to the texture film before you are ready for it. The act of removing the two sheets from each other could damage a paper drawing. These texture and shading films are, therefore, better-suited for use on polyester drawing film.

When the texture or shading film is placed, it must be burnished down. This process removes all air bubbles trapped under its surface. If they are left, they cause spots on drawing reproductions where the texture or shade becomes "out of focus" and dim. Bubbles are removed with a poke from a matte knife and the flat end of a burnisher. Go over the entire film area in this manner to fully adhere it to the drawing.

Stock

Texture and shading film come on sheets of 10-inches-by-14-inches with 9-inches-by-12-inches of usable space. Other sizes are available. If you are shading large areas, look for the largest sheet you can buy. These products always show a "line" where two sheets are joined together.

Diagonals

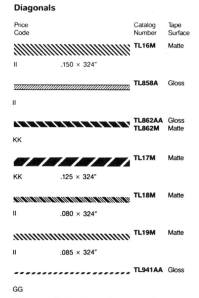

Price Code		Catalog Number	Tape Surface
II	.150 × 324″	TL16M	Matte
II		TL858A	Gloss
KK		TL862AA TL862M	Gloss Matte
KK	.125 × 324″	TL17M	Matte
II	.080 × 324″	TL18M	Matte
II	.085 × 324″	TL19M	Matte
GG		TL941AA	Gloss

Figure 4.19 Popular graphic tape patterns. (Popular graphic tape patterns courtesy of Chart Pak.)

GRAPHIC TAPE

Tape is available in a wide variety of repetitive patterns, widths, and colors for use on opaque drawings and charts as well as reproducible drawing sheets. Various patterns are excellent for graphically representing fire and smoke separations on a code plan, or outlining an area which is drawn in detail. On site plans, they can be used to show set-back lines, easements, utilities, overhead wires, and building outlines—anywhere you want a line to stand out above the rest of the drawing.

When using graphic tape on architectural drawings, always place it on the back side of the sheet. This is one of the reasons single-matte polyester film is used. The tapes can be applied and removed without affecting the drawing on the front surface. With graphic tape applied to the back surface, the front is available for continued drafting and erasure without being affected by the tape. Graphic tape will leave some residue of adhesive when removed. This can be easily removed with a solvent-based board cleaner. Again, if the tape had been applied to the front surface, it could not be possible to clean the remaining glue when changes are made. Another reason for placing tape on the back surface of the drawing is the reproduction process. Tape, placed on the top surface often adheres to the glass drum of a diazo printer. This will remove the tape from its intended location and cause images to occur where they are not wanted.

Limitations

Graphic tape can be used just about anywhere except the front surface of reproducible drawings.

Stock

Graphic tape comes in rolls 324-inches-long and is packaged in a plastic envelope to keep the adhesive from drying out.

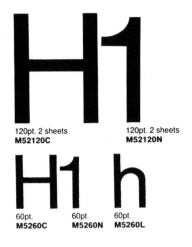

120pt. 2 sheets
M52120C

120pt. 2 sheets
M52120N

60pt.
M5260C

60pt.
M5260N

60pt.
M5260L

20pt.
M5220CL

20pt.
M5220N

18pt.
M5218CL
S5218CL

18pt.
M5218N
S5218N

Figure 4.20 Dry-transfer letter.
(Dry Transfer Lettering courtesy of
Chart Pak.)

DRY-TRANSFER LETTERS

Dry-transfer letters are matte-black letters on a polyester film sheet or strip. Letters are available in 15 point sizes from 8 pt to 96 pt and a large variety of type faces including Helvetica, Microgramma, and Times Roman. They are applied to a drawing by rubbing with a blunt instrument (burnisher) over the top surface of the carrier sheet until the entire letter adheres to the drawing surface. Be sure to rub over the entire letter, otherwise small pieces of the letter will not transfer to the drawing surface. After applying all letters in a series, use the backer sheet that comes with the letters and rub carefully over the entire area of the drawing.

If a mistake is made while transferring letters, there are a number of procedures for removing the error. On polyester drafting film, the incorrect letters can be removed by pressing a piece of clear cellophane tape over the letter. The letter will adhere to the tape when the tape is removed. If portions of the letter remain, as they often do, apply another piece of tape and remove it and repeat until the incorrect letter is removed. On paper where tape might tear the surface off when removed, consider scraping the incorrect letter(s) with a razor blade. If you are careful, the surface of the paper will still be workable. Another technique is to paint the error with rubber cement. When the cement is dry, the error will adhere to the cement as it is removed from the paper. If you are using transfer letters on nonreproducible paper or boards, it's a good idea to test these processes on some scraps before working on the final stock.

The use of transfer letters should be restricted to drawing and sheet titles and some drawing graphics where hand drawing of any kind will not occur. Once the tape cover is applied to the letters in place, it is very difficult to change any covered line work.

Stock

Transfer letters are available on sheets of polyester film, protected by a coated backer sheet to protect letters from sticking to anything until they are used. Each sheet has a random placement of letters, generally on the same sheet. Larger sizes have only caps or only smalls on a single sheet. Some letters are available in strips for use in drawing-sheet numbering.

Limitations

Transfer letters are all "right reading" and are, therefore, applied to the front surface of a drawing. There is a heat-resistive transfer letter made just for diazo reproduction, but it is likely the letters you are using are not made for hot-reproduction processes. This will soften the letters and make them come off the sheet. To prevent this, use a clear-film art tape to cover all letters. Dry-transfer letters may also dry out with age, causing them to crack and flake off the drawing. Again, cover them with clear-film tape to increase the drawing life.

MACHINE LETTERING

Machine lettering is produced on a graphic tape, in house, for a custom finished product. These machines let you prepare room titles, sheet titles and numbers, drawing titles and numbers, and any combination of words, numbers, and symbols you need. They are produced on a adhesive-backed polyester tape and come in numerous letter faces and point sizes from 10 pt to 36 pt. The newest electronic machines can even produce letters to 72 pt.

Machine-lettering systems are fast, provided the machine is located close to where you work. Letters can be made very quickly by typing entries on a keyboard. Letters are produced on polyester tape in the exact order as entered. The resulting tape has a protective backing which must be removed just prior to affixing the strip of letters to the drawing.

If an error is made, the entire ribbon of tape must be removed from the drawing. This is easy from polyester drawing film and vellum, but not so easy from bond papers and chart papers. Like any tape product, it will pull the surface off these papers. On polyester film, the tape will leave a residue after it is removed. This residue can be removed with a moistened vinyl eraser and a lot of rubbing. It is important to remove this residue, or it will begin to attract dirt and eventually leave a large dark mark on the drawing.

Figure 4.21 *Machine lettering.* (Machine Lettering courtesy of Kroy, Inc.)

Limitations

Machine lettering has some of the same problems inherent in transfer lettering. Some brands are applied to the front surface of a drawing and, therefore, are vulnerable to continued drafting processes and reproduction processes. The letters should be protected with a clear cellophane tape so they don't chip and flake. Do not place tape strips over other artwork. In many cases, the original artwork will need revisions, and this is not possible through mechanical lettering tape.

CONCLUSION

There are hundreds of other drafting tools and graphic supplies, too numerous to deal with here. As stated at the beginning of this chapter, make friends with a supplier and be a good customer. You can rely on this relationship to obtain sound advice on products and techniques for today and throughout your practice. Your supplier receives all the latest catalogs, and vendors are often looking for places to test their products. By knowing your dealer, you can stay on the leading edge of good production practices and equipment.

REPROGRAPHIC SYSTEMS

5

INTRODUCTION

This chapter is the second in a series of three chapters which deals with the tools and systems of architectural document production. In this chapter, the various processes of reproducing original documents are explored. This chapter deals with the products discussed in Chap. 4, and applies the art of reprographics. Chapter 6 will put these two elements together into a system for producing contract documents with the goal of greater economy of production and increased overall accuracy.

Original drawings are never sent out for bidding and construction. The reason is simple: There is no way we could produce all of the copies, by hand, that are required by agencies, owners, and contractors. We must rely on the reprographic industry to take our original work, copy it, and return it in good condition.

KNOW THE REPROGRAPHIC PROCESS

It is important that design professionals understand materials and systems used in reprographics in order to effectively manage a project. It is not possible for an architect to make correct decisions about drafting media and production systems without understanding how they are used in the reprographics shop. For example, an architect produced a site plan for his project which consisted of 21 layers of information. See Chap. 6 on overlay drafting. He had no idea of the limitations imposed by diazo overlay compositing. As a result, he could not obtain progress prints without utilizing expensive photoreprographic techniques. Had he understood the diazo process, he would have realized that only three layers of information can be successfully copied at any one time. This knowledge could have saved him a great deal of money in photoreprographics, not to mention the time necessary to coordinate 21 layers of information.

The first and most simple of these reprographic tools is the diazo copier. Most architecture and engineering offices have a tabletop diazo copier and an electrostatic dry copier to reproduce drawings, specifications, and correspondence. For the most basic of reproduction needs, this arrangement might work fine, but as you begin to use more advanced production techniques, you also need to use more advanced reproduction techniques. This all ties into the supplies and materials used in document production and reproduction.

Work with a Local Reprographics House

It is difficult for most architects to keep up with the latest techniques in reprographics. For this reason, it is advisable to do business with a reputable reprographics house. The reprographics manager can assist you in your reprographics needs and help keep you up-to-date about materials and techniques. This kind of support is especially useful when you find yourself in an awkward situation like being halfway through a project only to find out that you have been drawing base plans on unpunched drafting film (see "Overlay Drafting," Chap. 6). Reprographics managers can help you to avoid such mistakes in the first place. They can also help you to fix problems of this kind with minimum cost in time and money.

As I stated in Chap. 4, I work with Mike Murdock at Olympic Printing and Reprographics of Seattle. He has helped me keep current with reprographics techniques and equipment and to apply them in my practice. A person like Mike should be on every architectural project team.

Even with a knowledgeable reprographics manager on your team, it is important that the design professional have at least a rudimentary knowledge of reprographics processes. Following is a brief discussion of the most frequently used processes.

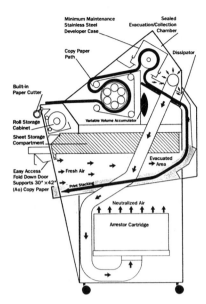

Figure 5.01 *Diazo process.*
(Diazo process illustration courtesy of
Diazit.)

DIAZO PRINTING

The most common request in drawing reproductions is for diazo-process blueline copies. It is the most economical process available for obtaining a right-reading positive copy of a large-format architectural or engineering drawing.

Diazo is a one-step contact-printing and developing process that can be performed in small tabletop machines without the need for special darkroom processing. Copies are made on light-sensitive paper or polyester film by exposure to ultraviolet light, and developed by either liquid or anhydrous ammonia. This simple process is the basis for a whole family of diazo-reproduction products and systems that have revolutionized the production of architectural drawings.

The Diazo Machine

The machine which delivers this one-step process consists of a feed board where the original drawing is placed, face up, over light-sensitive paper and together inserted. Drive belts carry the two sheets under a glass cylinder which houses an array of ultraviolet fluorescent lamps. These lamps (minimum of one and maximum of six) expose the light-sensitive paper except where the graphic images are. These images create a shadow on the print paper, protecting the light-sensitive coating. The drive belts then send the original to an outlet for removal by the operator. The print paper continues over and through an ammonia developer where the shadows left by the original graphics are developed, creating a new image. These machines have light-intensity switches, speed controls, and a forward/reverse switch. Ammonia gasses are taken away by mechanical exhaust ventilation or pass through an arrestor cartridge and neutralized.

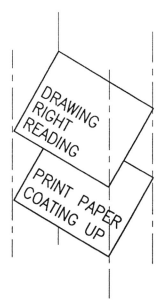

Figure 5.02 *Exposing a right-reading image.*

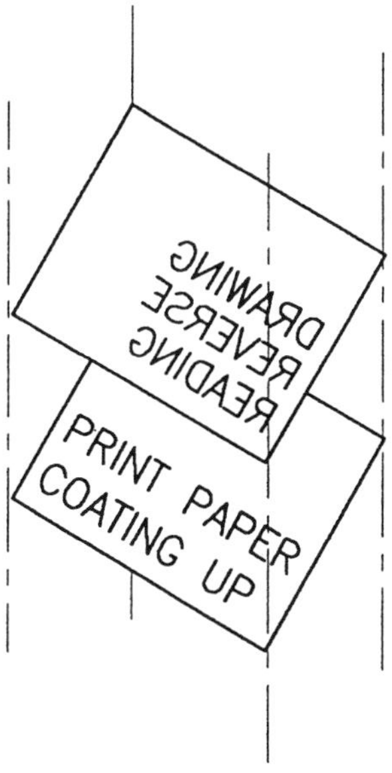

Right-Reading Versus Reverse-Reading

This reprographic process is really quite simple but a more thorough explanation of the exposure process will help you to determine how to produce drawings in your office for a long time to come.

Paper and film drafting products all have thickness, even if it is only .003 inches or 4 millimeters. This thickness will always keep the image of a right-reading drawing away from the print paper. Because of this distance, even as small as it is, light can pass under the graphic image and very slightly expose the light-sensitive paper. This results in a copy where all images are slightly fuzzy.

Because of this problem, a reproducible intermediate was developed. The original artwork is placed face-down against the photosensitive coating of the intermediate sheet and passed through the diazo printer. This results in a *reverse-reading* intermediate whose image is as clear as the original drawing, and sometimes even better. This intermediate is then used, image-side-down, to produce *right-reading* copies of higher quality to direct printing methods. The reason is simple: In both steps, the image is always tight against the light-sensitive copy paper, with no room for light to fade the images. This concept is important, and will become evident in Chap. 6 when we discuss overlay drafting.

Figure 5.03 *Exposing a reverse-reading image.*

Drawing "Creep"

Diazo printers all have a glass drum inside where the photo-sensitive paper is exposed. The round shape of this drum causes a "creep" in the reproduced image. Over the width of a 30-inch sheet, this "creep" will result in an image that is slightly larger than the original (about $1/8$-inch). Try this simple experiment. Take a plan drawing which fills most of the drawing sheet. Make a copy on your diazo copier, the way you usually do. Take a copy and place it on a flat surface. Place the original over it, aligning the image at the upper-left corner. If you run the print with the wide dimension of the drawing sheet into the diazo machine, the image will "grow" in the short dimension of the copy. If you run the print with the narrow dimension into the diazo machine, the image will "grow" in the long dimension of the sheet. This drawing "creep" caused the printed copy to be larger than the original drawing and, therefore, out of scale. If, for example, a reproducible intermediate (a paper or film copy used for producing finished copies) is run for making clearer copies, by the time the final copy is made, the drawing creep could make a plan $1/4''$ out-of-scale over its length. If you take one of these prints and attempt to work out a ceiling plan, for instance, you will find the rooms, as they are actually built, will appear to be inches smaller than they show on the drawing. It is actually the drawings which will be inches larger than the scale indicates them to be.

There is a solution to this problem. Make critical plan and building elevation copies on a machine that keeps the original and copy medium perfectly flat. The machine that does this is an offshoot of the light table, used for checking consultant's prints for years, and is called a *flat-bed printer.*

The Flat-Bed Printer

It works like this: The bottom part is an enclosed box. The top surface is a sheet of frosted glass which disperses the light from an array of ultraviolet fluorescent lamps. These lamps are controlled by a timer which regulates the exposure time to the light-sensitive print medium. There is a cover which has two main functions. First, it helps hold the drawing original and print medium tightly together. (Remember, any gaps will cause the print image to be fuzzy.) Second, it shields the operator from a blast of ultraviolet light. The base unit is equipped with a vacuum pump which removes all of the air from between the original drawing and the print medium. This pump is on a timer which automatically turns it on before the exposure begins.

There is also an array of fans on the cabinet's side. These fans merely cool the lamps and have nothing to do with the copy itself.

Figure 5.04 *The flat-bed printer.* (The flat-bed printer courtesy Design Mates, Inc.)

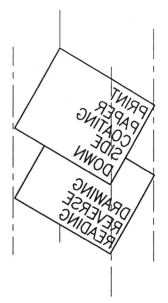

Figure 5.05 *Exposing a right-reading image.*

It takes about 20 seconds to produce a copy on the flat-bed printer. The procedure varies slightly, depending on whether the copy is to be right-reading or reverse-reading. For right-reading prints, place the original drawing face-down on the frosted-glass top. Place a sheet of light-sensitive print paper with the coated surface down over the original. Remove most of the air between these sheets with a table brush, and close the cover. Select the time the vacuum pump will run prior to beginning exposure. Select exposure time. (This takes practice to do correctly, so be prepared to trash a few copies until you determine the correct settings for the combination of drawing medium and print medium.) Press the start button. The machine does the rest.

When the exposure is complete, the lights are turned off, and air returns to the copy surface. The lid is still difficult to remove because a partial vacuum remains.

Caution: When copying unpunched drawings, the rush of air from opening the cover may separate the drawing and the print medium. This can be avoided by taping the print paper down on two corners prior to exposing.

Once the print medium is exposed, remove it and send it, coated side up, through the developer side of your diazo copier. The resulting print will be a true-to-scale right-reading copy. At this point, it is still not "top-quality" because there was still a thickness of drawing medium between the graphics and the print medium.

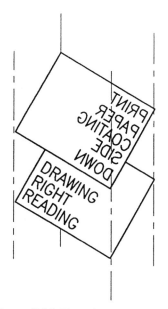

Figure 5.06 *Exposing a reverse-reading image.*

Use of Intermediates

For top-quality diazo reproductions, make a reverse-reading "intermediate" on reproducible paper or polyester film. (These products will be discussed later.) To make a reverse-reading intermediate, place the original face up on the frosted-glass top. Place the intermediate medium coated side down over the original drawing. This way, there is no space between the original line work and the light-sensitive coating of the intermediate. Tape these two sheets to the glass, top edge only, cover, evacuate the air in the printer, and expose the intermediate. The resulting intermediate, when developed, will be a reverse-image, and as clear as the original drawing.

This intermediate can now be used two ways. For progress printing, where scale accuracy is not an issue, print copies on a diazo copier. Place the intermediate image side down, as it is inserted with the light-sensitive copy paper. This will produce a copy of superior quality and with relative speed. If scaleable accuracy is needed, use the flat-bed printer. Place the intermediate, image side up on the frosted-glass top. Cover with copy paper, light-sensitive coating side down, and process as before. To prevent waste, always check the exposure schedule for the combination of original and copy medium that you are using. This process yields as many scaleable prints as you need, but it is slow. Allow enough time in your project schedule for making copies.

Making Composites from Overlay Drafting

The light table is an excellent machine for making a single copy or intermediate from multiple originals, as in overlay drafting, (See Chap. 6 for a discussion of overlay drafting.) All drawings are preregistered during the production process by using a seven-pin bar which aligns the seven holes at the top of the drawing stock. These holes are now used to realign the originals for copying. On the printer, special plastic "buttons" are used to align about two or three holes. This is enough to hold all layers during processing. Place the drawing originals face-down for prints and face-up for intermediates. Place the copy medium, light-sensitive side down over the originals aligning the two or three plastic buttons. Check the exposure schedule and make your copy. Develop as usual in the diazo machine. This is an economical and very quick way to make progress prints for overlay drafting.

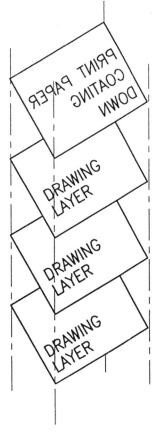

For Right Reading Copies

Place Drawings Face Down
And Cover With Copy Paper
Coating Side Down

For Reverse Reading Copies

Place Drawings Face Up
And Cover With Copy Paper
Coating Side Down

Figure 5.07 *Make copies from multiple originals.*

Making Screened Images

The diazo process and flat-bed printer are capable of producing screened images, either full-sheet or any part of a sheet. Start by purchasing a screen from your reprographic shop. Select a value between 30 percent and 50 percent. Place the screen right-side-up on a flat-bed printer. Place the copy medium, coated side down over the screen, and make a normal exposure according to the exposure schedule. Remove the copy medium and set it aside, face-down. Remove the screen and store it for future use. To finish the process, take the original drawing that you wish to screen, place it on the frosted surface of the flat-bed printer (face-down for right-reading prints and face-up for reverse-reading intermediates), place the partially exposed copy medium face-down over the original drawing, close the lid, and run the copy according to the exposure schedule. The resulting copy will be a screened image of the original drawing.

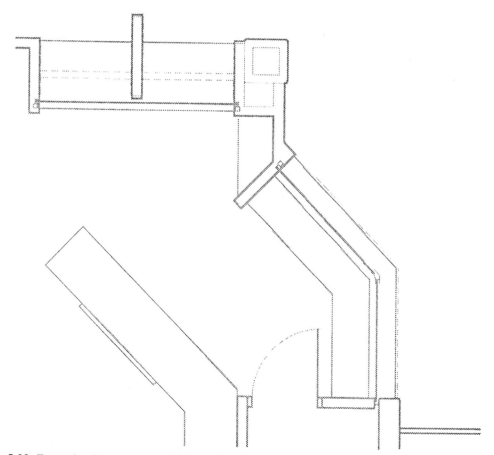

Figure 5.08 *Example of a screened print.*(Example of a screened image courtesy of Olympic Blueprint.)

Limitations

Diazo imaging is a fast and affordable way to obtain copies of architectural drawings for check printing and for distribution to agencies and contractors. However, it is not capable of providing clear copies for overlay drafting where there are more than two layers to be composited with a base sheet. The reason goes back to the earlier discussion about light going behind the drawn image and spoiling the quality of the copy. If this condition is noticeable for single-drawing original copies, imagine the problem when there is more than one original to be combined onto a composite copy. Each successive drawing layer becomes a little more hazy on the final copy. Even when making reverse-reading intermediates, the best copy is made from only two original layers. This can be very limiting in the production of overlay drafting (see Chap. 6). And, it is for this reason that drawing overlays are often limited to one. Diazo copying is also limited by the copy media it can use. If other finished products are desired, a different process must be used.

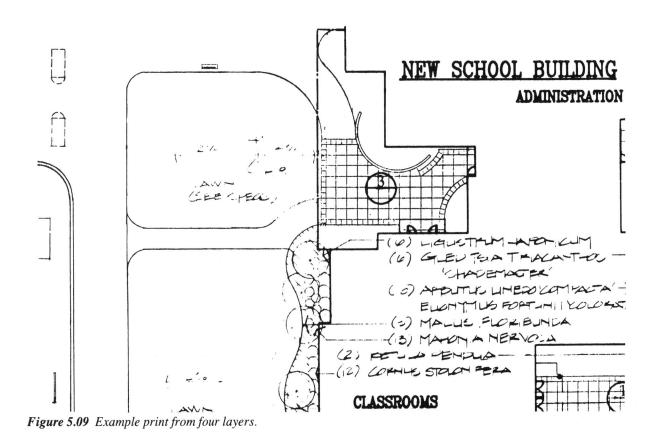

Figure 5.09 *Example print from four layers.*

Stock

The diazo process is well-known for blackline and blueline paper copies. These two familiar products are similar in many characteristics but differ in more than just color. Blueline is the standard for the industry. For the architect's print room, blueline is available in more varieties of speed, weight, and size than is any other diazo-copy media. Blackline paper offers most of the same choices as blueline except that fast speeds are not always available, and slower speeds may only be available in the lighter 20-pound-weight paper. Some sizes of specialty paper are also available only in blueline. Check with your local supplier for stock availabilities.

ANSI Y14.1−1980
STANDARD
DRAWING SHEET
SIZES

POPULAR
DRAWING SHEET
SIZES

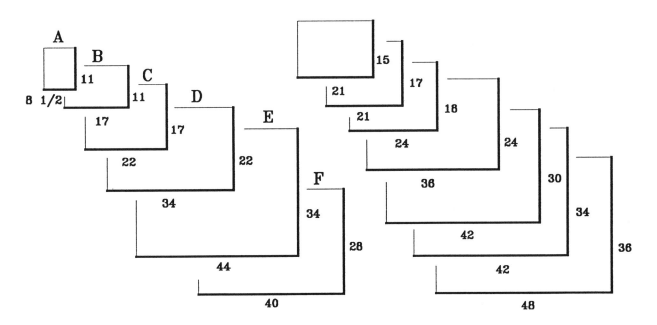

Figure 5.10 Standard precut sheet sizes.

Diazo copy paper is available in at least three speed ratings. These are medium speed for normal copying and greater clarity, fast speed for quicker printing but a lesser-quality copy, and superfast speed which runs about four times faster than the slower-speed paper. Diazo copy stock for office use is available in rolls of 50 yards-by-24 inches, 30 inches, 36 inches, and 42 inches. Paper weight is typically 20-pound. Other weights available include 24, 32, and sometimes 48 pounds. There is a third color of diazo copy paper available. Brownline comes in only the most common sheet sizes and in only one weight: 32 pound.

Intermediates and second originals are made on either vellum or polyester film. They provide a quick, inexpensive, reproducible copy of a drawing. These copies are used to make multiple blueline or blackline prints (as described earlier), and as records of the state of a drawing at any given time. Of the two materials, paper (or vellum) sepias are the least expensive. They have a fixed image which can be removed with the use of eradicating fluid, or there are some available with an erasable image, making it usable as an active second original. (Be careful with this one. No drawing can really have more than one original. Don't use this unless you need a workable, reproducible copy to spawn a totally new drawing.)

After paper sepia products comes the diazo-sensitized polyester films. These films come in many combinations of thickness, matte surfacing, erasability, image color, and sheet size. First, polyester film comes in 2-mil, 3-mil, and 4-mil thicknesses. The 2-mil thickness is the least-durable and is not recommended for use as a working second original, but because of its thinness, it makes better prints than do thicker films. The standard thickness for most purposes is 3-mil. Matte surfacing is usually on one surface with the diazo image on the other. The image can then be revised without damaging the drawing surface. Single-matte film makes a good second original and reproduces well. There is also a product with no matte used for best quality intermediates because there is no matte surface to cloud the image and slow it down. The 4-mil product is perfect for sharing base-layer information among the design team. Film is also rated by how fast it will reproduce when compared to each other. Standard speed is a base line which has no established value. The next speed is called "rapid," and relative to standard speed it is about twice as fast. The standard speed does not work well in tabletop diazo printers, as it is really designed for use in console machines. The rapid-speed film is better suited to office tabletop printers.

DIAZO PRINT MEDIA	SPEED			WEIGHT		THICKNESS		
	S	F	SF	20#	32#	.002	.003	.004
Blue Line Print Paper	o	o	o	o				
Black Line Print Paper	o	o		o				
Brown Line Print Paper	o				o			
Sepia Vellum	o			o				
Erasable Sepia	o			o				
Single Matte Diazo Film		o	o				o	
Double Matte Diazo Film		o				o	o	
Clear Diazo Film		o						o

Figure 5.11 Diazo print media.

Another variable in polyester film is the image color. It offers either traditional sepia color or black. The sepia color may be slightly faster when used as an intermediate because of the red pigments in the sepia image. The black color is often used because it looks more like ink. Of the two colors, blackline may be visually preferred, but it does not last as long under ultraviolet lights as does sepia color. As a working original, sepia color is best. Many of today's films are erasable with a wet vinyl eraser. Others are available with a fixed line which can be removed with an eradicating fluid. All these products come in standard sheet sizes and in rolls of 20 yards and 50 yards-by-30-inches, 36-inches, and 42-inches.

CONCLUSION

This review of the diazo-reproduction process and the many associated materials available for imaging was meant to give you a base of information so that you can discuss the alternatives with your reprographics manager. Learn to use most of the materials mentioned so that you can determine the best and the least-expensive media for any given situation.

PHOTOREPROGRAPHICS

There are many reprographics needs besides the one-to-one ratio provided by diazo copiers. Assume for a moment that you have drawn a building plan at ⅛″ scale. You are now considering detail plans at ¼″ scale, or even ½″ scale. You could regenerate the drawing by hand, but it is often easier to have an enlargement made photographically. You can have the image returned in so many formats that it may be difficult to decide which one you want. Here are just a few examples:

The image can be any size you want from 8½-inches-by-11-inches to 36-inches-by-48-inches, and larger on some cameras.

The enlargements can be printed on paper or polyester film.

The image can be right-reading or reverse-reading.

The image can be full-value or screened to one of about 10 values from 10 percent to 90 percent.

You can ask the reprographics manager to have the image "sited" wherever you want on the final sheet.

You can have multiple images composed onto a single sheet.

Photographic reproduction provides a wide variety of end products aimed at greater flexibility and increased economy in architectural-document production. These benefits are possible because of the highly accurate and very expensive cameras, developed and manufactured specifically for drawing reproduction.

The Reprographics Camera

These cameras consist of the following main parts: a subject holder which is either back-lighted for polyester film and vellum drawings or front-lighted for opaque original drawings, the lens which is often really an array of optical glass lenses of high quality and accurate settings, and film holder/negative holder which automatically places the film or negative for precise and accurate exposure and blow-back.

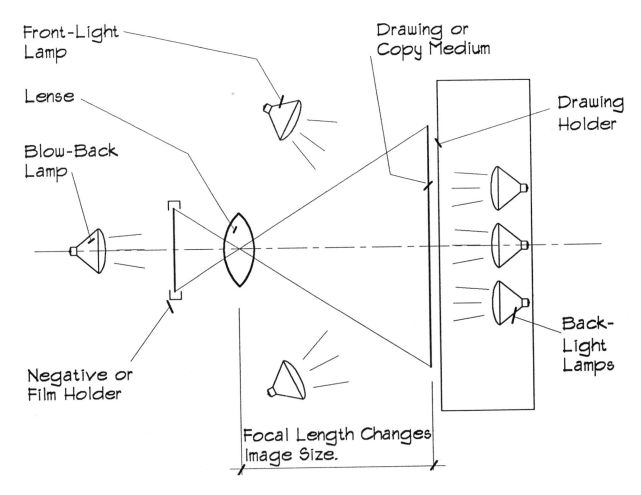

Figure 5.12 *Components of a camera.*

Copies are made by following these steps:

Expose A Negative

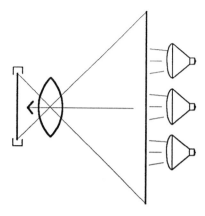

Figure 5.13 *Exposing a negative.*

The original drawing (assume an ink drawing on single-matte polyester film) is placed on the subject holder. (Proper placement is achieved by pin registration.)

The subject is back-lighted for optimum clarity of reproduction.

The camera is loaded with film. (Often 8½-inches-by-11-inches in size.) This film is also punched for accurate placement back into the camera after developing.

The camera is closed, and an exposure is made.

The film is processed and replaced back in the camera.

The original is removed from the subject holder, and a sheet of reproduction medium is placed on the pin registration.

The camera now acts as a projector to blow back the image on the new medium.

This medium is then developed.

The result is a custom as-you-ordered copy of the original drawing.

Blow-Back A Negative

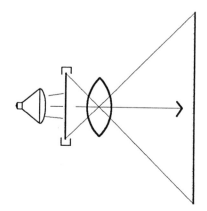

Figure 5.14 *Blow-back on light sensitive stock.*

Change Drawing Size

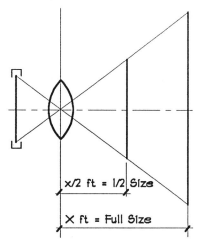

x/2 ft = 1/2 Size

X ft = Full Size

Figure 5.15 *Change focal length to change drawing scale.*

Photography is Versatile

The camera used for this process is very versatile. By changing the focal length, an image can be reduced or enlarged to any desired size. On some cameras this can be done to an accuracy of 0.001 inches. This accuracy is very important when copying and compositing originals from overlay drafting.

Unlike the diazo process, which copies all layers of a drawing in one exposure, photoreprographics copies one layer at a time. In this manner, each layer can have its own special treatment, such as screening. It also offers one very important quality: Every layer has the same clear, crisp reproduced image as the original, sometimes better. This allows the architect to combine more layers of information without sacrificing end quality.

Since the camera is very mobile and very accurate, it can be used to blow back images in precise locations. In this manner, these enlarged plans, that I mentioned earlier, can easily be located anywhere on a new drawing sheet. I have seen this ability used when the owner changed the scope of a project, deciding to place another addition to an existing building. The plans were already started for the project as originally conceived, but were too far over on the sheet to accommodate the new requirements. To solve the drawing problem, the drawings were lifted photographically and moved over by blow-back photography onto new polyester film. The new addition was drawn where it was supposed to be, and the remainder of the drawing was finished, by hand, as if no change was ever made.

Copies Are Only as Good as the Original

The photoreprographic process can provide a wide variety of imaging tasks, but the end product depends entirely on the quality of the original work. If the line work is poor, the resulting copy may be less than desirable. Line work should all be from the same medium. Ink is best, since it is formulated especially for the reprographic process. Plastic lead works well with ink for lettering. It also produces a dense black image for reproduction. Graphite and clay lead should be avoided when photographic reproduction is used because the camera can not be adjusted for both ink and graphite images on the same drawing. Graphite is very reflective, causing the reproduced image to fade and become weak in relation to ink or plastic lead. As a result, the line work drawn in graphite might fade away. For best quality photoreprographics, always use ink for drafting.

Photoreprographics can be used for a variety of needs. The following is a brief discussion of some of them.

Restoration

Restoration of old and weak originals is possible through photoreprographics. Drawings produced years ago on vellum with pencil may be yellow, faded with use, and wrinkled from abuse. These problems make the drawing difficult to reproduce and to obtain clear copies. Once, the only way to fix this problem was to redraft the entire drawing. With photoreprographics, the original is shot on a high-contrast negative and blown back on high-contrast drawing film to enhance poor line quality. Negatives can be doctored before blow back to remove unwanted images like coffee spills and creases in the drawing paper. The new original can be used for many functions that the old original could not be used for like microfilming, overlay drafting, and screened images.

Photo-Drafting

Photo-drafting uses actual photographs blown-back on polyester drafting film as a means of expressing a situation more clearly. I worked on one hotel restoration project where we needed to show the contractor where selective elements of the building facade needed repair. We hired a photographer to take pictures of each elevation with a special camera which records all lines in true elevation, not in perspective. These photographs were blown back at exactly ⅛″ scale and screened so that the gray tones would reproduce. This was combined with the office standard title block on wash-off polyester drafting film to create a drawing without actually drawing a line. These photo drawings were then marked to show the extent of repair work. This technique can be used with any photograph for which a negative exists. It works very well on interior remodeling jobs where a lot of areas need restoration. Consider incorporating photo-drafting wherever a photograph can be used to more clearly or more quickly define the work.

Paste-Up

Paste-up drawings (see Chap. 6) are assemblies of drawings which are combined photographically. A good example is a sheet which contains all of the GFRC (*Glass Fiber Reinforced Concrete*) panel details for a building. I like to draw all of my details on 8½-inch-by-11-inch detail sheets (see Chap. 1), and at the end of the job, combine them through photoreprographics. I have clear-film reproductions made of each detail. Then, I cut out the portion needed for the final drawing and tape it to a carrier sheet. This carrier sheet, with all of its contents, is photographed and blown back onto draftable, wash-off polyester film. The film can be prepunched for overlay work. Selected images can be screened for desired effect. The original paste-up is no longer needed, but should be retained until the project is completed.

Opaquing

Opaquing is a process of revising drawings right on the negative rather than on the original. Like in drawing restoration where unwanted wrinkles are edited out, any unwanted information can be selectively removed from the intermediate negative. The process involves "painting" an opaquing fluid over the area to be deleted. For large areas, a red film or tape is placed over the negative in the area to be edited out. This process can easily save time over erasing the same information from a blow-back wash-off. It will also likely save the drawing surface which can easily be damaged from large-scale erasures. Opaquing can be used when you don't want to change an original drawing. For example, you have a file of drawings that are close to being reusable on other projects, but some of the notes, or elevation marks, or dimensions are too specific to be reused. The drawing can both be reused and saved without changing the original through the process of opaquing. Have an intermediate negative shot of the original drawing, and return the original to the file. Edit the negative through opaquing, and have it blown back on wash-off polyester film. New information can now be added to finish the drawing.

Overlay Drafting

Overlay drafting (see Chap. 6) involves separating the information on a drawing into layers. The *base sheet* is usually the architectural plan without notes, dimension, and detail references. This clean base becomes the foundation for mechanical and electrical "overlays." These are registered by using a seven-pin bar and prepunched drawing film. Since there is no drawing sheet which shows the finished project, composites must be made, and for final documents, photoreprographics is the best method. It produces a new combined original that is as good as a hand-drawn original using ink.

Each layer of a drawing is photographed individually. Some layers may be screened. This is common for architectural base sheets under consultant layers. Selected layers are then blown back, one at a time, onto photosensitized polyester film until a desired drawing is produced. Since there is only one architectural background original, and everyone needs it to draw their respective layers of information, it can be copied several times from one negative and distributed to everyone who needs it.

Direct Duplications

Direct duplications are often required. For example, at the close of each phase of work, i.e., schematics, design development, and contract documents, it may be necessary to make a second original to use as an intermediate for printing the progress sets. This intermediate then becomes a reproducible record of the phase completion which is durable and permanent. Direct duplication is also used when the project is "out-of-town" and the cost of printing and mailing from your location would be prohibitive. Second originals can be made photographically on fixed-image polyester film and mailed for reproduction near the site. Fixed-image polyester film is used to prevent others from altering your work.

Scaleable Images

Scaleable images are used to reduce drafting time and take advantage of the camera instead of the pencil. There are two examples. When you work on an apartment complex, the "typical" plan is often drawn at $1/4'' = 1'-0''$ scale. You might also need a $1/16''$-scale overall composite plan to show how all of the units fit together to make up a building. Have the $1/4''$ plan photographically reduced and copied as many times as necessary. Cut and tape these copies into the desired plan configuration and have this photoreproduced onto your standard title and border drawing sheet. When the drawing comes back, fill in corridor end walls and building core, and you have a plan.

Wall sections can be produced in similar fashion. Draw details of roof edge, window openings, and other exterior elements at 1 ½″ scale of 3″ scale. Have them photographically reduced. Cut and tape them to a carrier sheet so that they begin to show a wall section. Have the paste-up photographed and blown back on wash-off polyester film. When the drawing is returned, connect the vertical lines, add notes, dimensions, and detail references, and the drawing is done. Drawing reduction by photography reduces line thickness as well as the whole image. If you use this process, consider using thicker line weights for drawings at 1 ½″ and 3″ scale.

Screen Tint

Halftones (called *screen tints* in the reprographic house) are screened images in which some percentage of the drafting image is removed through a photographic screening process. The resulting blow-back is gray in appearance. When new line work is added, it is very black in contrast and, therefore, reads much stronger. Screen tints are often used when remodeling selective areas of an existing building. If you have accurate "as-built" drawings, have them screened and use them as the background to draft over with new work. The new line work and notes will read very strong, making it obvious to the contractor, the scope of new work. The most popular screen reduces the original line work by 50 percent of the original value. Consider screens which produce as low as 30 percent of original. Less than 30 percent may not copy if the diazo speed is too slow.

Microfilm

Microfilm has been used for years by governments and large architecture and engineering firms as a means of reducing archive space. When you decide the time has come to archive a project, consider microfilm. The product is now good enough to produce workable original blow-backs, should the need arise in the future.

Comment

These brief descriptions of some of the features of photoreprographics are intended to whet your appetite for additional information. Your reprographics house will be more than willing to explain any of these in more detail. The goal is to reduce production costs overall, and to increase the accuracy of what is shown.

Stock

The materials used in photoreproduction are many, with each having its own specific purpose. For reproducible drawing originals there are basically two. First is the conventional fixed-line photosensitized polyester drawing film. This film can be ordered right-reading if the finished photoreproduction will be drafted on and reverse-reading if this is only an intermediate for reproduction. This product comes with a drafting surface for additional work and clear for intermediates. If the image must be revised, do not attempt to erase it. The image can only be removed by a three-step eradicator process. Even this, if not done properly, may leave permanent marks on the drawing. This conventional stock is best-suited to direct duplications for archive and for printing.

Wash-off photosensitized polyester film is the second and most popular of the photographic originals because it can be erased. This product comes with image right-reading or reverse-reading, and with a drawing surface on one or both sides. The main purpose of this product is as a draftable original which will likely require some changes. For full-value line work, order right-reading images so that they and the new work are on the same surface. This saves flipping the drawing over to make erasures. For screen tints, order reverse-reading images. These images are more likely to remain unchanged, and you don't want them where they can be erased along with new drafting. Also, they will print better on the back surface.

Photoreprographics is the primary method for producing final composites from overlay drafting. The end product, whether fixed-line, wash-off, or intermediate clear film, is the finest quality reproducible available today. They are permanent, nonyellowing, durable copies that can be used and stored for years.

ELECTROSTATIC COPIERS

Electrostatic dry copiers have become as important in drawing production as the pencil. These machines make copies of almost any original and produce a copy of clear, crisp images which often rival the original. The typical copier uses three sizes of paper. These are *letter* (8½-inches-by-11-inches), *legal* (8½-inches-by-14-inches), and *ledger* (11-inches-by-17-inches). These machines are most often thought of for copying the typewritten page, but they are very versatile for copying drawings.

When details are drawn on the standard "A"-size detail form, they can be easily copied for publication on the electrostatic dry copier. This process is far less expensive than full-sheet diazo processing or "A"-size offset printing, making the office copier the medium of choice for all small-format drawing copying.

Some specific uses for the office copier follow. These are just some of the ways in which the machine can help you save time and money during project design and drawing development. Be sure that the copier is set to make precise one-to-one ratio copies.

Using the Office Copier

When you are developing a series of details that have repetitive elements, like window conditions, hollow-metal frames, and roofing flashing, it is very handy to draw the repetitive element once and copy it. This copy can then be rotated, reversed, or moved at will to any new place on a drawing sheet and traced. Reverse images are made by copying a drawing on clear film, and recopying the film image backwards onto paper. Copies can also be made so that the repeat image is always in the same location on successive details. This works well for roof-flashing conditions. The top of flashing should have a constant height anyway, so it can be shown graphically by being in the same location on all details. Since the flashing is the repeat element, draw it once, and copy it onto drawing stock. Then, continue to draw the remainder of the detail. If a repeat image is subject to change, however, consider hand tracing of the repeat item. Use the copier to produce the image that you slip under the drawing stock. The reason for hand tracing is, dry-copier images are not erasable. If the repeat element changes, you would need to redraw the entire drawing.

The office copier makes excellent copies for cut-and-paste detail sheets. To do this, take the drafted detail or a detail from file and make either an opaque copy for photoreprographics or a clear one for diazo copying. Cut away any unwanted image, and tape the remainder to a carrier sheet (see more on cut-and-paste drafting in Chap. 6). Reproduce the carrier sheet by the appropriate method. Take the new full-sheet original and finish the drafting manually.

The office copier can be used with adhesive-backed reprographic film. Copies are made and affixed to a drawing like sheet appliqués. Any image that is able to be copied can be used. A popular example for many years has been the schedule form. Make an appliqué by first copying the desired image on clear film. Turn the intermediate film over, and copy a reverse image on adhesive-backed reprographic film. The new appliqué can then be placed on the back side of a drawing sheet.

The technology which makes the office dry copier possible is known as *xerography*. This technology has evolved from simple letter-size copies at one-to-one size ratio, to scaleable copiers which are capable of reducing a letter-size drawing to the size of a postage stamp and enlarging a postage-stamp image up to 11-inches-by-17-inches. Other copy machines can now make full-size drawings up to 36-inches-wide by any length. This means that xerographic technology can now be used in lieu of diazo copying for making prints. One of the primary advantages of xerographic large-format copying is that reproducible copies can be made on vellum and polyester film from opaque originals. The only other way to do this is by photoreprographics. These machines also make reduced copies to half-size and enlargements to 200 percent of original. The cost of xerographic large-format copying is becoming more compatible with diazo imaging. This should make xerography a viable addition to many architectural and engineering reproduction departments.

Laser Imaging

Another advance in dry copying is *laser imaging*. Through this process, color copying is now available with features previously available only in paint programs for personal computers. Images can be copied directly from photographs, or the machine can be used as a color laser printer for images from television, video cameras, and computers. The features seem endless, and for at least today, so is the price. Color laser-imaging machines are only affordable by the large reprographic houses, but luckily, their capabilities are there for you to use.

TAKE A TOUR THROUGH A REPROGRAPHICS HOUSE

A reprographics house will likely have all of the equipment and stock discussed in this chapter and more. These things are arranged to give you, as their client, the most efficient and highest-quality reprographics product and service available. At Olympic Reprographics, their facility is located on two floors. The main floor has camera rooms, darkrooms, and CAD plotting rooms. Olympic has four cameras, each made for specific reprographics purposes.

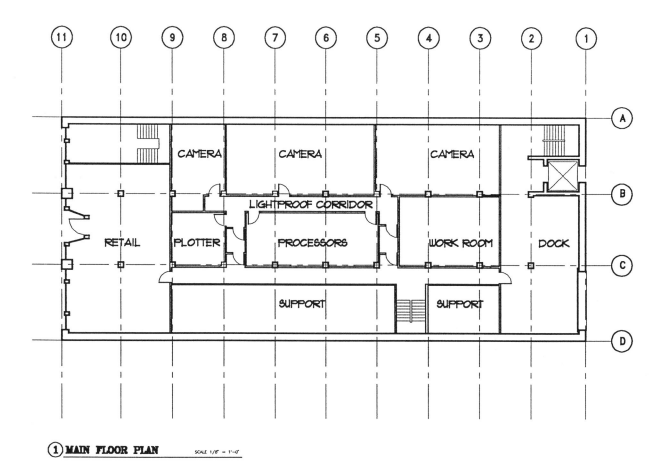

Figure 5.16 *First floor plan with camera rooms.* (First floor plan with camera rooms courtesy of Olympic Blueprint Company.)

Upstairs is the diazo room and the electrostatic copy room. There are four commercial diazo copiers, a vacuum frame and a Roll VAC which works like a vacuum frame except that it uses vertical film mounting and is much faster. Following the diazo area is a xerographic copy center which includes a color laser machine that can reproduce images from videotape. Next is the offset print room where another camera is dedicated to 8½-inch-by-11-inch printing. In the center of the floor is a bindery where copies from all processes are bound and wrapped for return to the customer.

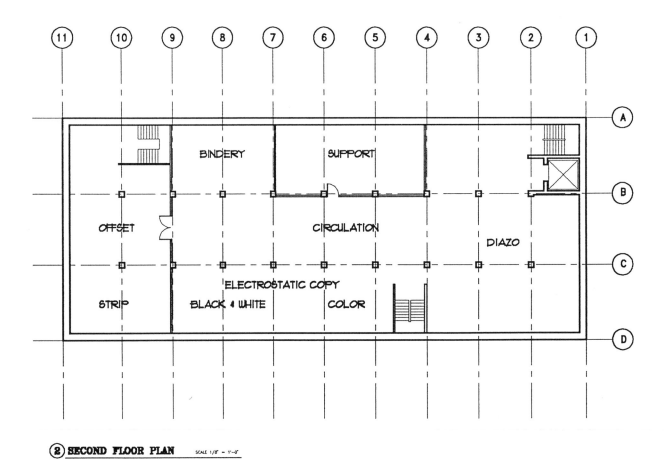

② **SECOND FLOOR PLAN**   SCALE 1/8" = 1'-0"

Figure 5.17 *Upper floor plan with diazo, offset and electrostatic copiers.* (Upper floor with diazo, offset and electrostatic copiers, courtesy of Olympic Blueprint Company.)

CONCLUSION

The reprographics systems available to an architect do far more than just make blueline copies. They have developed in sophistication, speed, and quality so that they now perform a variety of functions including photoreprographics, drawing restoration, and photo-drafting. These reprographic systems now allow the architect to use production systems (see Chap. 6) like cut-and-paste and overlay drafting to speed up the design process and to make it more profitable.

PRODUCTION SYSTEMS

6

INTRODUCTION

This chapter is the third and final chapter in a series dealing with the actual production of architectural drawings. Chapter 4 explained graphic media, how they are used, and some of their limitations. Chapter 5 explained the reprographic processes most often used to produce and copy drawings. This chapter will work with the background of information from Chaps. 4 and 5 to describe systems for producing drawings. It is the goal of these three chapters to give you enough information to help you to reduce the time it takes to produce the documents for a building, and to improve the accuracy of the drawing set before it leaves the architect's office.

Every architect has experienced problems while producing project documents. Some of the more common ones are the following:

Too much time spent redrawing familiar details, job after job

Mistakes made when redrawing or tracing existing details

Incomplete consultant coordination

Incomplete architectural coordination

These problems, the time they consume, and the money they waste eat away at the very foundation of an architect's business. They also cause client dissatisfaction and give the architect a reputation within the construction industry for poor drawings. Many of these problems can be avoided. By having set methods for producing drawings and specifications, much of the waste and many of the mistakes caused by poor production methods can be eliminated. Below is a quick list of the areas where standardization and systems can help:

Basic drawing production through a systems approach

Cut-and-paste drafting techniques

Overlay drafting techniques

CAD (*Computer-Aided Drafting*)

CAP (*Computer-Aided Production*)

Having systems in each of these areas will reduce wasted production time and improve drawing accuracy. The result will be a more profitable business, happier clients, and a reputation for top-quality drawings.

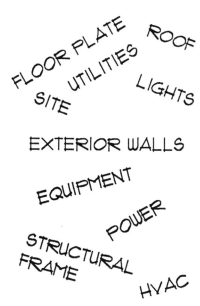

Figure 6.01 *Common parts of any building set.*

PRODUCTION SYSTEMS

A *production system* can be defined as an established way or method for producing documents which can be used regardless of the kind of building that is being designed. This is important. The building type has little to do with the way the drawings are produced. This is one of the basic tenants of a production system. *There are certainly differences in buildings but it is the parts which are common to all structures that makes a production system work.*

Consider the Options

There are several production systems being used and promoted around the country. These include ConDoc™, which is a professional-development program by the American Institute of Architects, and GUIDELINES™, which is a production system for architects by a private "self-help" organization of the same name. Each of these systems promotes concepts of production similar to the one proposed in this book. They differ, however, in the manner in which information is stored and retrieved. This book recommends their literature for its general content so that better decisions can be made in setting up a production system.

Any full-scale production system addresses the process of drawing development from basic drafting through some rather sophisticated reprographic techniques. Much of this has already been discussed in Chap. 4 and Chap. 5. It is the very important work which is done in between these two areas which will be addressed now.

SYSTEMS PRODUCTION

Systems production is aimed at finding and using the common elements in document production and has as its guide three main elements. These are the following:

Prepare reusable elements.

Maintain a file system so these elements can be stored and retrieved.

Assemble the reusable elements into new projects.

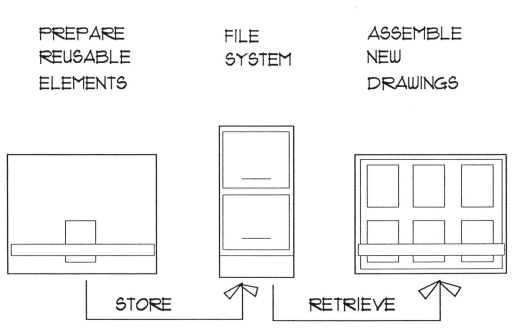

Figure 6.02 *Systems production.*

PREPARING REUSABLE ELEMENTS

One of the problems in architectural practice is the repeat drafting being done unnecessarily. Door details are a primary example. The standard 16-gauge hollow-metal frame in drywall has been hand-drafted a million times or more. This replication of efforts costs money, and in those million times has, all too often, been the cause of drafting errors.

That relatively simple hollow-metal detail does have some important elements. Unfortunately, the very simplicity of the detail often causes architects to hurry through it, without thinking, making costly mistakes. I know of a project where the hollow metal had to be rejected and reordered because the throat size was incorrectly noted on the details. All the door frames in the project had to be replaced at the architect's expense. That one simple mistake cost the architect a great deal of money.

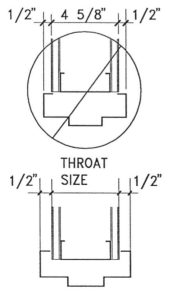

This frame will never fit over the partition.

The actual partition size can be as much as 4 3/4'.

Throat size should be 1/8' larger than the nominal partition size.

See SDI-100 for more information

Figure 6.03 Simple mistakes can cost a lot of money.

The redrawing of repeat elements costs an architectural office both time and money, and places the architect at risk over details that should have been standard for decades. The answer to this problem is to prepare drawings and other elements which can be reused.

A Drawing System Uses Reusable Elements

The previous example about hollow-metal door frames is only one area where reusable elements will help. Following is a list of areas where the development of reusable elements is necessary to establish a systems production.

- ### Standard drawing format

- ### Standard details

- ### Standard schedule formats

- ### Standard notations

In systems production, each of these must be developed as a unit and to complement the others.

STANDARD DRAWING FORMAT

Drawings are usually produced on large drawing sheets. The space within the title block is often considered to be free territory to use as the architect pleases to draw plans, elevations, sections, and details. This is fine if the idea is to continue redrawing everything from scratch. This does not work, however, for standardizing production. The drawing sheet needs order.

Organize the Basic Drawing Sheet

Details are drawn on 8½-inch-by-11-inch drawing sheets. This makes them easy to handle, store, copy, and reuse. These attributes make the 8½-inch-by-11-inch sheet the root for all drawing sheets. Large-size drawing-sheet order begins with the smallest drawing-sheet size, 8½-inch-by-11-inch. This size can be divided into two spaces 7¼-inches-wide-by-4½-inches-high, which become the standard module for all drawing production. By making this module standard, every drawing, schedule, and note sheet will fit within a predefined space or some multiple of this standard module.

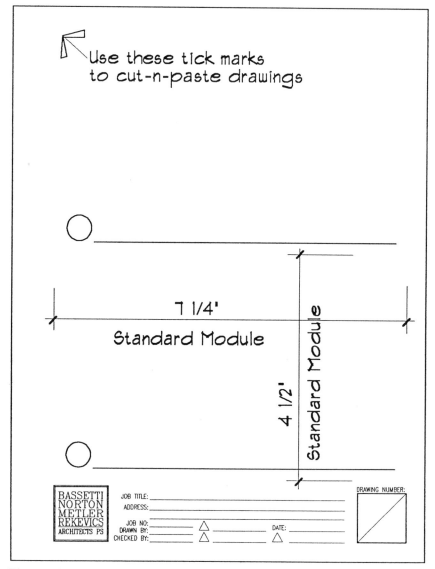

Figure 6.04 The standard detail sheet.

Evaluate the Three Popular Sizes

The three most popular drawing-sheet sizes are as follows:

24-inches-by-36-inches

30-inches-by-42-inches

34-inches-by-42-inches

See Chap. 1 for a basic description of drawing sheets and their title blocks.

Each of these popular drawing sheets can be arrayed with the standard 7¼-inch-by-4½-inch drawing module. The 24-inch-by-36-inch drawing sheet contains 20 drawing modules, the 30-inch-by-42-inch drawing sheet contains 30 drawing modules, and the 34-inch-by-42-inch drawing sheet contains 35 drawing modules. If maximum number of drawing modules was the criteria for selecting sheet size, the 34-inch-by-42-inch drawing sheet would easily be selected. But there are other reasons for selecting the drawing-sheet size, and all must be considered together.

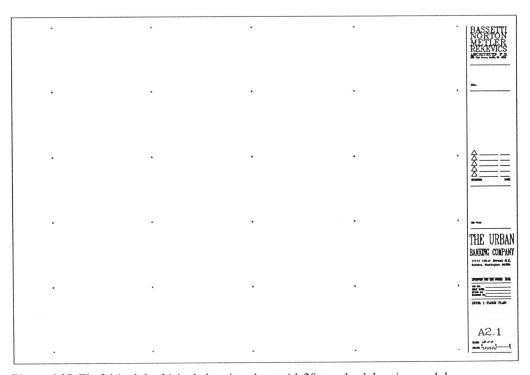

Figure 6.05 *The 24-inch-by-36-inch drawing sheet with 20 standard drawing modules.*

Figure 6.06 *The 30-inch-by-42-inch drawing sheet with 30 standard drawing modules.*

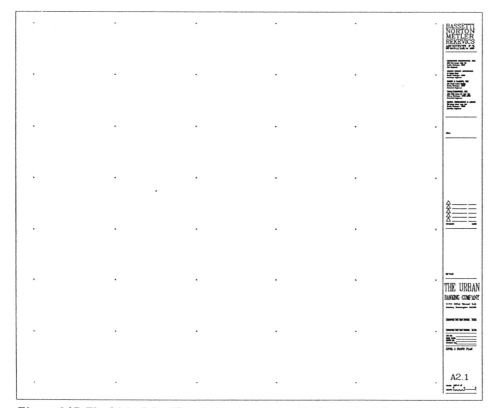

Figure 6.07 *The 34-inch-by-42-inch drawing sheet with 35 standard drawing modules.*

Test the Economy of Each Drawing Sheet

Notice that on the 8½-inch-by-11-inch sheet, there are two standard drawing modules, one above the other. These two modules are often treated as one when drawing details because the detail consumes the whole space.

Of the three most popular drawing-sheet sizes only one has multiples of two modules high: the 30-inch-by-42-inch size. This means it is the only sheet size that can carry details that consume two modules each. The other sheet sizes have an odd number of drawing modules. Details will be more difficult to arrange on these sheets because the top or bottom row of modules may not be filled. Consider this potential for wasted space when selecting sheet size based on floor-plan space requirements.

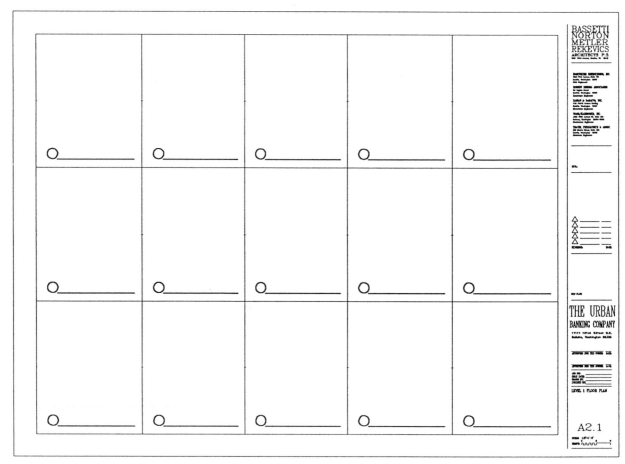

Figure 6.08 *Array of full-size details on a 30-inch-by-42-inch sheet.*

Fit the Building Plans on the Drawing Sheet

Another consideration to note when selecting sheet size is that every drawing should fit on either one, two, or some multiple of the standard drawing module. For example, if a floor plan fills a 24-inch-by-36-inch drawing sheet, it will fill all 20 standard modules. If the same floor plan is drawn on a 30-inch-by-42-inch drawing sheet, it still occupies only 20 standard modules, leaving 10 modules available for other information.

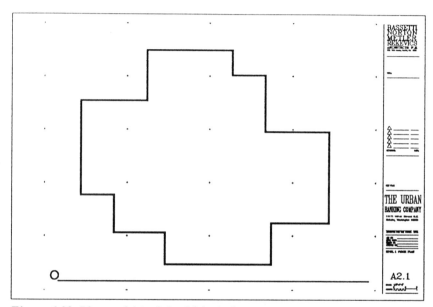

Figure 6.09 *Plan on 24-inch-by-36-inch sheet.*

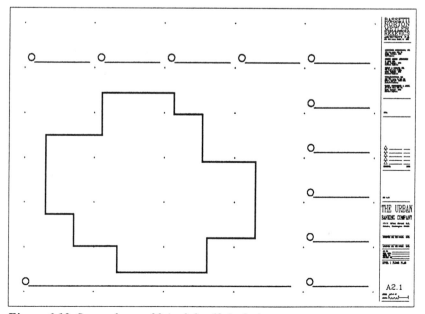

Figure 6.10 *Same plan on 30-inch-by-42-inch sheet.*

Layout an Interior Elevation Sheet

This standard module also works for producing interior elevations. Use the vertical dimension only as the guide for establishing rows of elevations. Since the length of an elevation is a product of the room size, it cannot be regulated by the drawing module width. A 30-inch-by-42-inch drawing sheet will carry six rows of interior elevations.

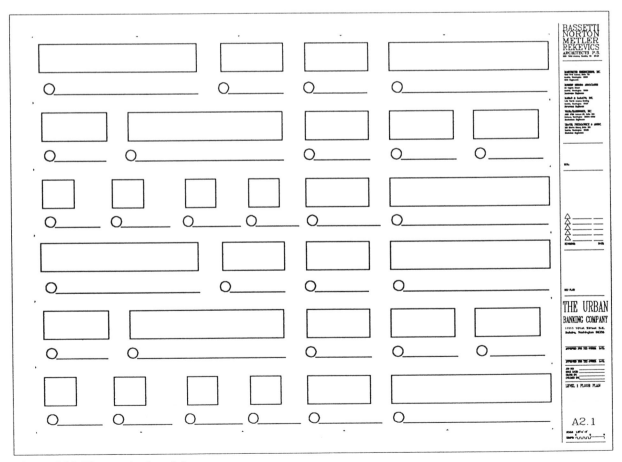

Figure 6.11 *The standard module.*

The Standard Module

The *standard module* is divided into two areas. The area on the left is for graphics, and the area on the right is for notes. This simple order is the root of an entire production system. The standard drawing module of 7¼-inches-by-4½-inches sets the size and arrangement of all drawings. It brings reusability to details and order to large-size drawings and is the first reusable element of any production system.

Figure 6.12 The standard drawing module.

STANDARD DETAILS

The next reusable element of production systems is the *standard detail.* Standard details are used, like all production system elements, to cut production time and to reduce errors. The door-frame example used earlier would not have occurred if standard hollow-metal details had been available.

Another architect that I know was pressed for time and sketched hollow-metal details for a project on 8½-inch-by-11-inch drawing sheets, but did not have the time to draft them in pen and ink. They were eventually reproduced from the original freehand sketches, and published for bidding and construction. Because time was short, the sketches were done quickly and rather messily. As it happens, these were done correctly, but when time is short and the pressure is on, mistakes can often be made. It would have been so much easier, neater, and faster to have selected the needed details from a Library of Standards. This would also have reduced the possibility of error.

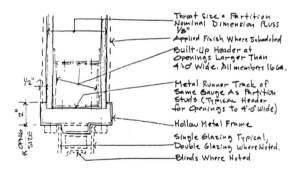

Figure 6.13 *This is an example of a hastily produced freehand detail.*

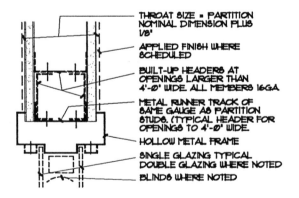

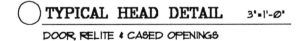

Figure 6.14 *The same detail from a Library of Standards.*

Select and Review Details to Become Standards

Details as simple as door frames should be complete and ready to use. This would improve project economy by reducing production time. The place to store these details is in *the Library of Standards* (see "Maintain a File System" later in this chapter). When details are selected for the Library of Standards, they must undergo a quality-control review. In the case of the frames that were too small, standard details could have saved the architect a lot of headaches. The details would have already been checked for accuracy, and could simply have been retrieved from the library and added to the detail set.

Details under consideration for the Library of Standards can be selected from previous projects. They must be tested and known to work. Then, make prints or copies of these details on 8½-inch-by-11-inch paper for easy handling. As details are drawn for new projects, use the standard detail sheet and add them to the collection. Do not allow details that are known to leak or fail in any way to get into the library collection. If there is any doubt about the quality of a detail, have it reviewed by either the quality-control manager or the lead technical architect.

When beginning a set of details, start with a blank 8½-inch-by-11-inch detail sheet. Compose the graphic element on the left of the page in the space provided. Place notes on the right side.

Some standard details cannot be completed in one sitting before they are filed. Some information often needs to be added which is job-specific. This is alright. Do not take a drawing too far, or it will have to be erased.

Details can be either drawn by hand and copied for reuse or drawn by CAD and replotted when reused. The manual system requires the copy to be essentially correct and needing no erasure. This is at least in part because the image from the office copier cannot be erased. Information can be added, but not easily removed. CAD-produced details can be edited very easily to suit new conditions, since the information is electronic until it is plotted on the final drawing. This makes editing easy, but also increases the chance for making mistakes. For a detail to be truly "standard," it must remain relatively unchanged from file to job.

STANDARD SCHEDULES

Chapter 3 discusses the kind of information that goes into many typical architectural schedules. This chapter will deal with the way these schedules fit into production systems.

This portion of Chap. 6 is dealing with the reusable elements of production systems. Schedules themselves are seldom reusable once the spaces are filled in. However, their format is reusable. As noted earlier, every drawing and schedule should fit the standard drawing module or some multiple of it. Door and Opening Schedules and Room Finish Schedules are too wide to fit on one standard drawing module, so they are designed to fit in two modules (14½-inches-wide). They then run in lengths of 4½ inches, 9 inches, 13½ inches, 18 inches, etc. down the drawing sheet.

NO.	OPENING SIZE	RATING	ACOUSTICS	DOOR TYPE	TYPE	FRAME HEAD	JAMB	SILL	HARDWARE H	G	C	S	DP	T	W	REMARKS	REV
																	△
																	△
																	△
																	△
																	△
																	△
																	△
																	△
																	△
																	△
																	△
																	△
																	△
																	△
																	△
																	△
																	△
																	△
																	△
																	△
																	△
																	△
																	△
																	△
																	△
																	△
																	△

DOOR & OPENING SCHEDULE

SCHED1.DWG

Figure 6.15 *Many schedules are two-spaces-wide.*

Schedule forms can be produced manually by drafting the line work and adding the written data. Or, they can be produced by CAD and the schedule line work automatically arrayed. Schedules can also be produced completely by computer with the output copied to the drawings. There is more on this later under CAD and CAP.

Following are some ways to speed up the schedule-making process. Once the basic column headings have been determined and the form sized to fit the drawing module, try these techniques:

Example 1: For small projects with about 20 rooms or doors, have reverse-reading self-sticky images made on the office copier and apply them to the back surface of the drawing film. Then simply fill in the blanks. (Use some spray fixative on the copy image to keep the line work from cracking and falling off.)

Example 2: For large projects, schedules should be produced with the help of computers. Schedules can be published in book form by taking a computer printout and reducing it to fit the standard office drawing sheet. Schedules can also be plotted directly on drawing sheets by using CAD. (There will be more on CAD and CAP later in this chapter.)

STANDARD NOTATIONS

As discussed in Chap. 1, general notes are used to preface any grouping of information in the drawing set. This is done to help the contractor read the drawings or to find information within the drawing set. Notes of this nature are often typical, at least in basic content, so they can be produced by word processor and stored for future use. The word processor should be formatted to print these notes on the standard drawing module. When it is time to recall these notes, the basic content should not be changed. Change only such things as drawing-reference numbers. To change the main content of any standard note is to rewrite it. This negates all the protection built into the note when it was selected as a standard.

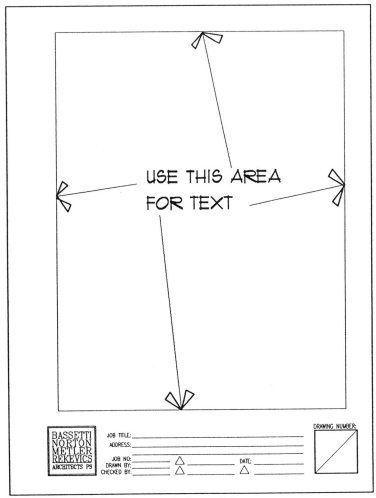

Figure 6.16 *Produce notes to fit the module.*

MISCELLANEOUS STANDARDS

There are numerous items which are used repeatedly in architecture that do not fit in the previous categories of standards. The following is a list of just three of these:

Cover sheet

Abbreviations list

Symbols lists and legends

These items are generally described in Chap. 1 because they are graphic standards. Since they are standards, they also belong here with production standards. The abbreviations and symbols lists are produced on 8½-inch-by-11-inch drawing sheets. They are then stored in the Library of Standards for future use.

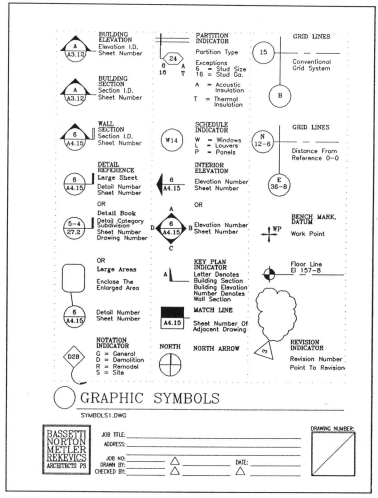

Figure 6.17 Graphic symbols on a standard detail sheet.

DRAWING STANDARDS ON A COVERSHEET

These two standards are also produced in large-sheet format as a cover sheet. A photograph is taken of the full-size cover-sheet master and the negative is filed in the Library of Standards for blow-back later.

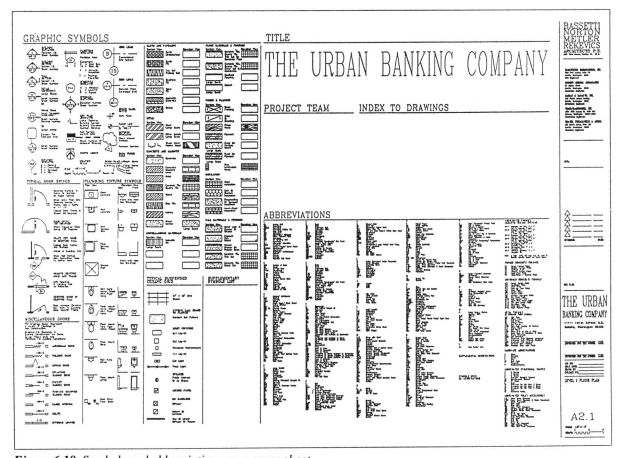

Figure 6.18 *Symbols and abbreviations on a cover sheet.*

STANDARDS SELECTION

This chapter has been dealing with the reusable elements of production systems. These are:

Standard drawing sheet and drawing module.

Standard details

Standard schedule forms

Standard notes

Before any single item is selected for the Library of Standards, it is first reviewed and approved for use by the quality-control manager or the lead technical architect. Today's production systems, especially the electronic ones, make it very easy to store and reuse these elements as well as to edit them. For this reason, great care must be taken to maintain the characteristics that made these elements standards. The quality-control manager or lead technical architect must review all changes to previously selected standards.

The next element of systems production is the system used to file and retrieve the standard elements.

MAINTAIN A FILING SYSTEM

Standard drawing elements aren't much use if you can't find them and find them quickly. A file-and-retrieval system must be developed which provides for this and more. The file system must have the following features:

It must be fast and easy to use.

It must provide an accurate method for finding information.

It must be easy to understand.

It must be comprehensive.

The basic file system proposed here is called the Library of Standards. It is a comprehensive assembly of all standard production items from preprinted drawing sheets to ceiling details. Because of the wide variety of size and forms of the library elements, they may not all be stored in the same place.

Large-format drawing standards like "slip sheets" with standard drawing modules and "cover sheets" with symbols and abbreviations should be kept in flat files with blank drawing stock in each drafting area. Standard schedule forms in appliqué format should be kept with other graphic products like film appliqués and graphic tape. Usable quantities of small-sized preprinted standard items like abbreviations lists and graphic symbols should also be stored with appliqués and other graphic products.

Negatives of large-sheet standards are saved for reuse in new work. These should be kept in an 8½-inch-by-11-inch loose-leaf binder titled *Negatives.*

Filing the Standard Details

The remaining contents of the Library of Standards are filed by using the drawing-index-numbering system also described in Chap. 7. This index system is used to publish details in book form, and it works equally well as a file storage and retrieval system for the Library of Standards. The basic filing system is as follows:

LIBRARY OF STANDARDS INDEX

0 GENERAL

0-1	General Architectural Drawing Index
0-1S	General Structural Drawing Index
0-1M	General Mechanical Drawing Index
0-1E	General Electrical Drawing Index
0-2	General Notes
0-3	Graphic Standards
0-4	Abbreviations
0-5	Legal Descriptions
0-6	Project Identification
0-7	Temporary Facilities

1 SITE WORK

1-1	Site Drainage (CB, YD)
1-2	Pavement and Walks
1-3	Site Furniture and Improvements
1-4	Landscaping

2 EXTERIOR CONDITIONS

2-1	Concrete (CIP, Precast)
2-2	Masonry (CMU, Brick, Stone)
2-3	Misc. Metal Louver Schedules (see also category 6).
2-4	Wood, Stucco, Plastic
2-5	EQJ/Exterior
2-6	Prefab Panel Systems (GFRC, Brick)

3 MOISTURE PROTECTION

3-1	Membrane Roofing
3-2	Roof Accessories (Hatches)
3-3	EQJ/Roof
3-4	Membranes (Sheet, Fluid)
3-5	Shingles (Wood, Tile, Composition)
3-6	Metal Roofing

4 WINDOWS

4-0   Schedules
4-1   Metal Sash
4-2   Storefront, Curtain wall
4-3   Wood Sash
4-4   Miscellaneous (Atrium)

5 DOORS, RELITES, AND OTHER OPENINGS

5-0   Schedules
5-1   Hollow Metal
5-2   Wood
5-3   Specialties (Roll-up, Overhead)

6 STAIRS AND RAILINGS

6-0   Schedules
6-1   Concrete Stairs
6-2   Metal Stairs
6-3   Wood Stairs
6-4   Ladders, Catwalks
6-5   Railings

7 INTERIOR CONDITIONS

7-0   Schedules
7-1   Floor and Base
7-2   Walls, Partitions, Furring
7-3   Ceilings, Soffits
7-4   EQJ/Interior

8 SPECIALTIES

8-1   Toilet Partitions and Accessories
8-2   Conveying Systems (Elevators)
8-3   Operable Walls, Folding Doors
8-4   Misc. Specialties
       Curtain Tracks
       Drapery Tracks
       IV Tracks
       Chalk and Tack Boards
       Dock Facilities
       Chutes
       Access Panels

9 CASEWORK AND MILLWORK

9-0   Schedules
9-1   PLAN Casework
9-2   Metal Casework
9-3   Hardwood Casework
9-4   Demountable Casework
9-5   Millwork

10 EQUIPMENT

10-0   Schedules
10-1   Fixed Equipment

Using the File System

When standards are selected for the loose-leaf binders, they are inserted, first-come-first-in. They are identified by category and subcategory numbers such as 5-1 for hollow metal, followed by a number representing the next consecutively numbered detail in the hollow metal subcategory. This number will always refer to that specific and unique detail or schedule form until it is removed from the library. This number can then be used to retrieve a drawing for future use.

When CAD and CAP are used to store, retrieve, and print or plot a drawing or schedule, computer files on floppy disk are kept in a desktop cabinet next to the bound hard copies. In this manner, a detail can be selected by looking through the hard copies and then copied into a computer for production by copying the soft copy on floppy disk. For people using DOS machines, the file system works equally well as the computer file address. If you are using UNIX or other operating systems, you need to consult your user's manuals for file designation.

ASSEMBLY OF DOCUMENTS

Once the drawings, schedules, drawing modules, and miscellaneous standards are established and properly stored away in the Library of Standards, they can be retrieved and assembled into new drawings. The process of assembling reusable drawing elements is the final step in this series of events. Now is the time to apply the concepts of reprographics to produce new drawings of high quality. With skill, the final drawings produced through the assembly process will look at least as good as first-time originals. This quality is a necessary goal because the end product of assembly must still be reproduced at the time of final printing.

Production Systems

The system used to assemble these reusable elements varies with the kind of drawing element and the manner in which it is to be reproduced. Some of the systems to consider include the following:

Appliqués

Cut-and-paste

Overlay drafting

CAD

CAP

The remainder of this chapter will cover each of these assembly/drawing systems.

Recap

First, this is a good time to recap the production systems process as discussed so far. The reasons for any production system are basically two. They are the following:

Increase accuracy of the documents

Increase profits

The basic elements of systems production are:

Prepare reusable elements

Develop a file system so those elements can be stored and retrieved

Assemble the reusable elements into new projects through various techniques including cut-and-paste, appliqués, overlay drafting, CAD, and CAP.

CARTOONING

Along with systems production, there is a need to see what the end drawing product will look like before it is drawn. The process for seeing and, therefore, planning the future drawing set is called *cartooning*. Small-size drawing sheets are used to sketch the basic contents of the final large-size sheets.

Start by making reproductions of the standard drawing sheet at one-fourth the actual size on 8½-inch-by-11-inch bond. Include the standard drawing module printed in nonreproducible blue. A stock of these forms should be on hand so that drawing sheets can be cartooned at any time. Obtain a supply of these cartoon sheets and begin planning the job.

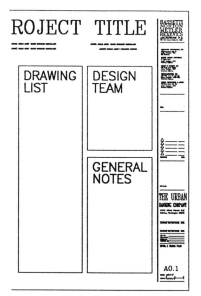

Figure 6.19 Index to drawings.

Figure 6.20 Prepare cartoon sheets on 8¹/₂-by-11-inch bond.

Use Cartoons to Begin the Index to Drawings

Number and title each drawing sheet (see Chap. 7) just as it will appear in the Index to Drawings. (In fact, the Index to Drawings can be started along with this cartoon set.) Draw a very rough cartoon of the main elements on each sheet. For example, draw a rough outline to represent the floor plan. Be sure that it fits within the standard modules with room for dimensions, grid lines, and the drawing title.

Cartoons Are Drawn to Scale

To assist in accuracy, use this rule of thumb: For $\frac{1}{4}$-inch scale actual plans, use $\frac{1}{16}$-inch scale cartoon plans. For $\frac{1}{8}$-inch scale actual plans, use $\frac{1}{32}$-inch scale cartoon plans. For sections drawn at $\frac{1}{2}$-inch actual scale, use $\frac{1}{8}$-inch cartoon scale. For details drawn at 3-inch actual scale, use $\frac{3}{4}$-inch cartoon scale, and for details drawn at $1\frac{1}{2}$-inch actual scale, use $\frac{5}{8}$-inch cartoon scale. Any cartoon scale can be determined by dividing the actual drawing scale by four.

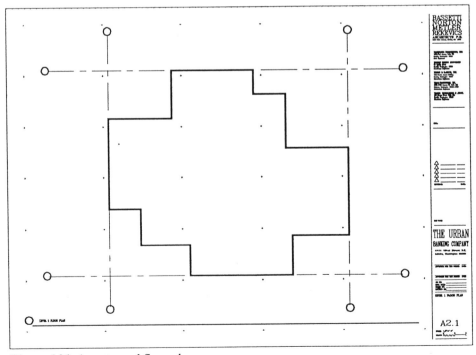

Figure 6.21 A cartooned floor plan.

Estimate the Number of Drawings

Draw cartoons for building elevations and building sections, wall sections, schedule sheets, and detail sheets. Estimate the size of all schedules so that they are accurately planned for on the drawing sheet. The same holds for details. It is only an estimate, but try to be as accurate as possible in setting the number of details needed for any area of the drawings. An accuracy of plus or minus 30 percent is good at this phase.

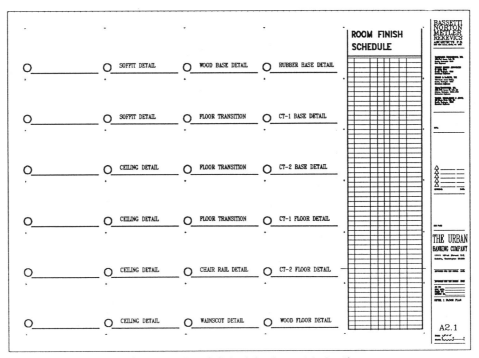

Figure 6.22 *Cartoon of a Room Finish Schedule sheet with details.*

Sheet Order Starts at Top Right

Arrange all drawings (except interior elevations) *from top to bottom* and *from right to left*. In this manner, any empty space will occur at the binding edge where it is difficult to see anyway. For interior elevations, arrange left to right then top to bottom. It is too difficult to visualize a series of wall elevations when they are numbered backwards.

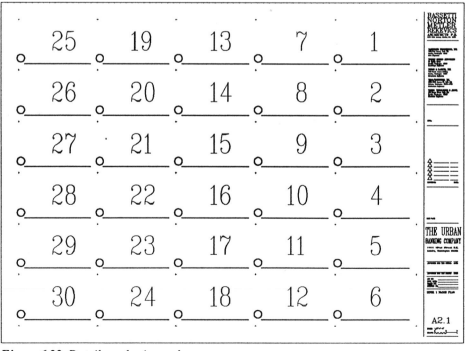

Figure 6.23 *Detail numbering order.*

Keep Cartoons Up-To-Date

As detail sheets are drawn, a cartoon for each should be drawn (on 8½-inch-by-11-inch sheets) on the cartoon set and the appropriate space titled and numbered to represent the actual detail. This will be used when the details are taped up for the final drawing.

Have the cartoon set bound in loose-leaf form so that the pages can be added and taken out for copying. Keep the bound set available for the project staff to see and use so that they know where information will go.

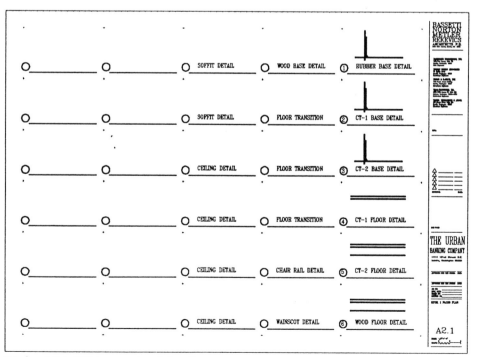

Figure 6.24 *Fill in completed details.*

SETTING UP FOR PRODUCTION SYSTEMS

The systems that follow are used to assemble the reusable standard elements discussed above, but they also have a broader purpose in architectural-document production. With these systems, every drawing can be thought of as an individual element of graphics which is portable and potentially reusable. Whether it is selected as a "standard" or not makes no difference in the way the drawing is produced. All drawings are, from this point on, produced as if they are going to become standards. This means, for starters, that every drawing should be drawn with ink on polyester film for best results in reprographics. Next, every drawing must fit the standard drawing module or a multiple of that module. If it does not fit this format, it will not integrate into a standard drawing sheet, and will, therefore, be a one-time-only drawing. In order to obtain maximum efficiency from the drawings, they must be drawn to fit the system. To assist in the requirement, it is helpful to have the right equipment.

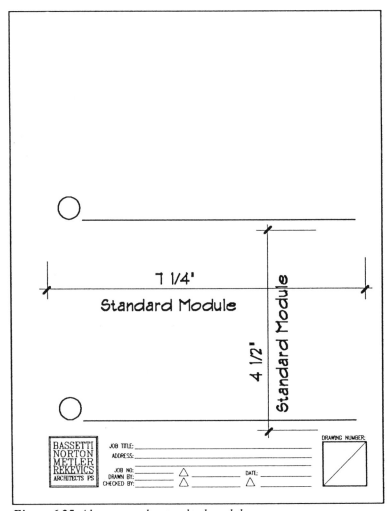

Figure 6.25 *Always use the standard module.*

Setting Up the Drawing Table

Every drawing table should be set up with the following production aids. First, obtain a drawing-board cover, a sheet of ⅛-inch gridded polyester film (see Chap. 4). This sheet must be of sufficient size to cover the entire drawing surface. This sheet of grids is "squared" with the drafting straightedge and then taped down to the table. Next, take a polyester film copy of the standard office drafting sheet with the drawing module arrayed on it. (See reusable elements.) Slip this sheet under the gridded board cover and square it with the straightedge. Set a "pin bar" along the top drawing edge of the slip sheet, align it with the seven holes, and tape it down. From that point on, produce all large-sheet drawings on pin-registered drafting film using the slip sheet and gridded board cover as guides to drawing location. When drawing on 8½-inch-by-11-inch stock, align the preprinted "tick marks" with any two drawing modules that are located at a comfortable working position. By using these basic tools, the requirement for standardizing drawing layout will be all but automatic.

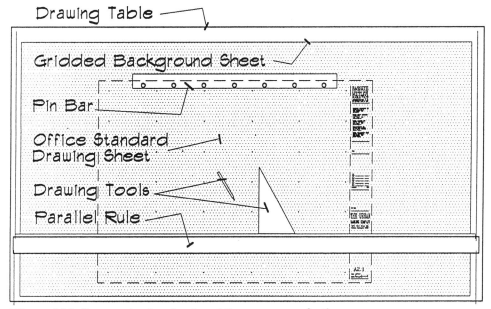

Figure 6.26 *Prepare the drawing board for systems production.*

CUT-AND-PASTE

With the drawing table set up for the coordinated production of reusable products, it is time to discuss the other elements of production systems. The first of these is *cut-and-paste drafting* (also known as *scissors drafting*).

Cut-and-paste drafting is a system for reusing existing drawings (and new drawings), with the aid of reprographics, to create a new drawing. Consider this example. Assume the task is to prepare a large-sheet drawing for doors and other wall openings. The cartoon shows a schedule in the upper-right corner of the sheet, followed by door-type schedule, hardware-location detail, frame-type schedule, and the details. This is a small project with about 25 doors, so that they all fit on the preprinted door-and-opening schedule form located by the graphic supplies. Obtain a copy and begin to fill it in by entering door numbers and door sizes. Set it aside and begin the other drawings, all on new 8½-inch-by-11-inch sheets using pen and ink.

Collect Elements from the Library of Standards

Go to the Library of Standards and select the appropriate details. Make copies on polyester film and edit to suit the project. Put them aside with the partially completed schedule form. Now, draw the door-type schedule and door-frame schedule on individual 8½-inch-by-11-inch drawing sheets.

Tape Up Individual Drawing Elements

When all of the 8½-inch-by-11-inch drawings are about 90 percent finished, and details have been selected from the Library of Standards, it is time to make copies to cut-and-paste for the drawing. First, make a right-reading diazo blackline "slick" of the office standard title block. Place this "slick" over the pin bar which is taped to the drawing table. This becomes the carrier sheet for the cut-out images. Make reverse-reading diazo slicks of each drawing and the schedule. Cut off the excess border on each and arrange them on the carrier sheet as shown on the cartoon set. Tape each drawing to the carrier sheet by using small amounts of clear-film tape along the top edge only. This will allow rolling the drawing (top-to-bottom) and also allow diazo copying in a rotary diazo machine. Progress prints are best made on a flat-bed printer. This ensures that the taped drawings will not come loose during copying. It also produces a copy of exactly the same size as the original.

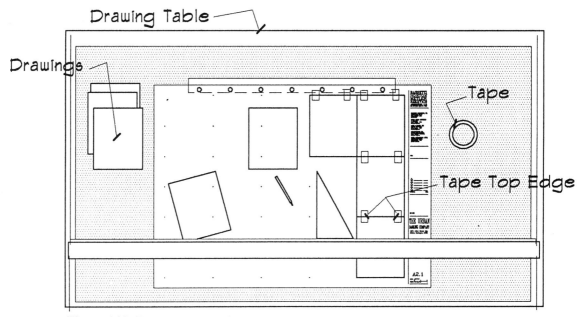

Figure 6.27 *Tape reverse-reading images to carrier sheet.*

Diazo Reproduction Possibilities

The diazo process can be used to make a variety of reproducibles. These include the following:

Blueline prints for progress checking

Diazo slicks for reproducing many prints

Diazo reproducible drawing film for economic originals when additional drafting is required

In place of diazo slick copies of each drawing, copies can be made on overhead-projector film on the office copier. These make acceptable images, but are not quite as true to line weight as is the diazo process. Economy and accuracy must be weighed to determine the method used.

Final Reproduction

When it is time to make a final original from the tape-up, send the artwork to a reprographics house for photographing and blowback on wash-off polyester film. Order a right-reading copy so that it can be "worked" without flipping the drawing sheet over to make erasures. Finish drawing on the front surface prior to final printing.

With the photographic original drawing in hand, it is not necessary to keep the taped-up drawing. The slick details, etc., can be removed from the carrier sheet to be reused on another project. The carrier sheet itself can be reused again and again.

This cut-and-tape process can be used for any drawing sheet, although it works best for details and small schedules. Use it anywhere small elements can be drafted on 8½-inch-by-11-inch drawing sheets, copied, and taped to a carrier sheet. In this manner, large drawing sheets are not produced until the last minute, when it is most certain that their contents will be correct and coordinated. In the interim, the individual 8½-inch-by-11-inch drawings can more easily and affordably be copied on the office copier.

OVERLAY DRAFTING

For many years, consulting engineers redrafted architect's plans for backgrounds over which they produced their own work. This meant that as many as five people redrafted the architect's floor plan for every engineering discipline including plumbing, heating and ventilating, power, and lighting. The duplication of efforts not only wasted a great deal of time, but opened the door for errors made during the redrafting process. A new system had to be developed which allowed the engineers to use a common architectural base plan and draw only their own work. That system is *overlay drafting*.

In overlay drafting, the architect splits the information shown on plan drawings so that the common elements can be shared with the consulting engineers. Then each person produces unique drawing layers representing his or her discipline. The drawings are eventually put together by either diazo or photoreprographics.

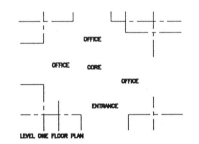

Architectural Note Sheet

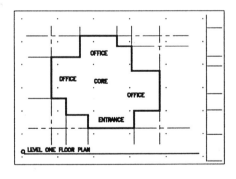

Composite Architectural Floor Plan

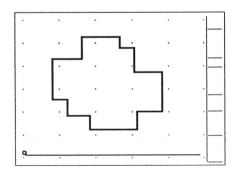

Architectural Base PLan

Figure 6.28 The layering concept.

Set Up the Drawing Surface

The drawing-table set-up described for general systems production is perfect for overlay drafting. The pin bar is used to maintain registration between drawing layers as well as aligning all work over the standard-module slip sheet. The pin bar can hold about four or five sheets of polyester drawing film depending on the thickness. (4-mil single-matte film for base sheets and 4-mil-high translucency film for overlays.) It is inadvisable to have more than one base and two overlays. The system becomes difficult to manage and reproduce when there are more than two overlays.

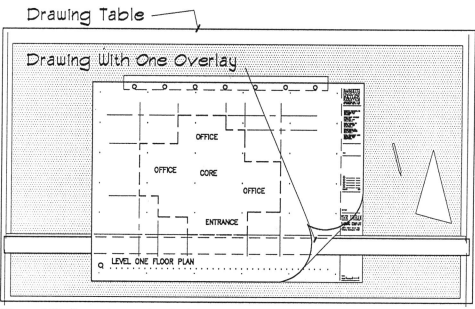

Figure 6.29 *Set-up for overlay drafting.*

Layering Information

The first task in overlay drafting is to determine what information to draw on the base sheet and what information to place on the overlays. Keep in mind the reason for doing overlay drafting: to separate and share common data from unique data. This will make your decisions easier. And don't overdo it. The system falls apart when layers are added "just because they can be."

The Architectural Base Plan

As an example for discussion, let's use the Urban Banking Company, a fictitious project, with banking on the first floor. The plans will be drawn using overlay techniques.

The first step in the overlay process after cartooning the sheet contents is to plan the contents of each drawing layer.

The base sheet for the floor plans will contain these basic elements.

###	Building grid
###	Walls
###	Windows
###	Doors
###	Base cabinets
###	Floor curbs, drains
###	Plumbing fixtures

These elements are common to all trades and will appear as a part of every discipline's plans, i.e., mechanical and electrical. Each of these consulting engineers will then draw a unique layer of information for each construction trade.

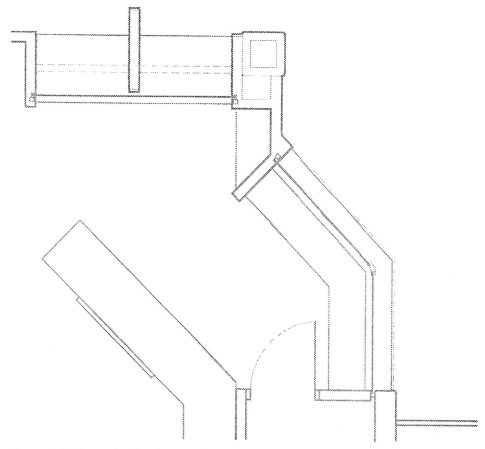

Figure 6.30 *Example of architectural base without overlay information.*

The Architectural Overlay

The architectural floor plan is completed on one or more "overlays". For most work, a single overlay is all that is needed. It will contain the following items:

Poché

Room names and numbers

Door and other opening numbers

Dimensions

Reference indicators

Notes

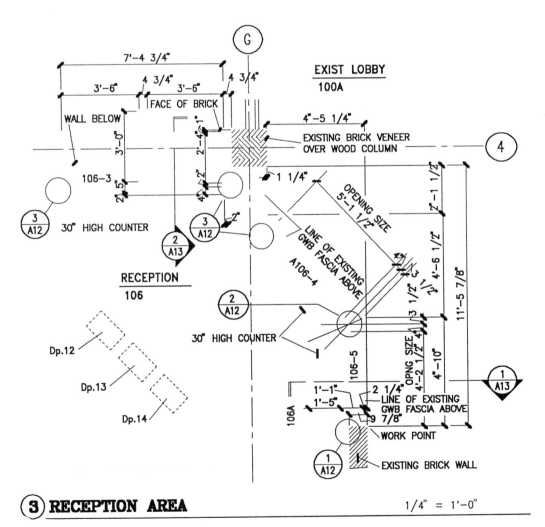

③ RECEPTION AREA   1/4" = 1'-0"

Figure 6.31 *The architectural overlay.*

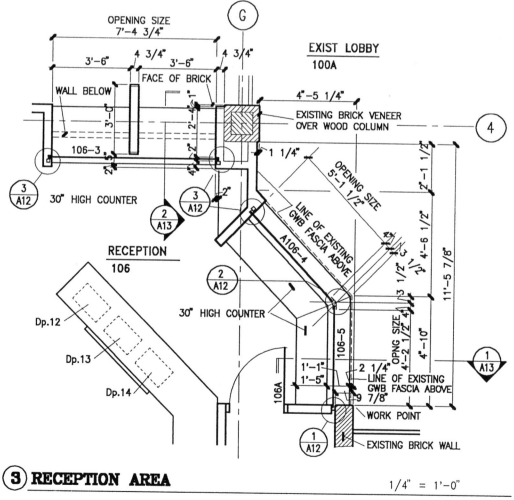

3 **RECEPTION AREA**   1/4" = 1'-0"

Figure 6.32 *Example of composited architectural plan from base and overlay drawings.*

The Material/Finish Overlay

These two drawings, base plan and one overlay, can normally carry all of the data that makes up the final drawing. In fact, one of the few reasons to have more than one overlay is because there is a need to ultimately produce a different drawing with the same base sheet. This could happen when making color and material plans. The base-plan drawing is unchanged. The unique layer would contain the following items:

Room name and number

Graphics to denote materials and colors

Legend

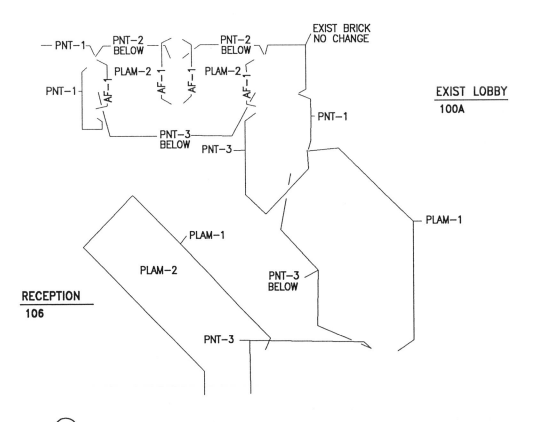

③ **RECEPTION AREA**

Figure 6.33 Example color and material plan layer.

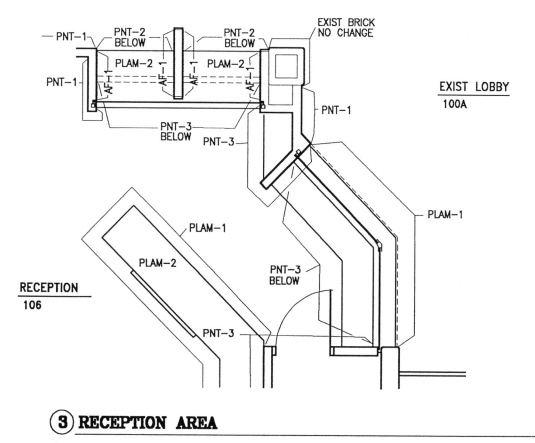

③ RECEPTION AREA

Figure 6.34 *Example of composited architectural base with color, material overlay.*

The Presentation Overlay

The architectural base can also be used for a presentation drawing. Since all of the working notes and poché are on the architectural overlay, a new overlay can be made just for presentation. Room titles can be made with mechanical lettering machines and walls blackened to add contrast. Appliqués can be added to shade and poché areas for special effects.

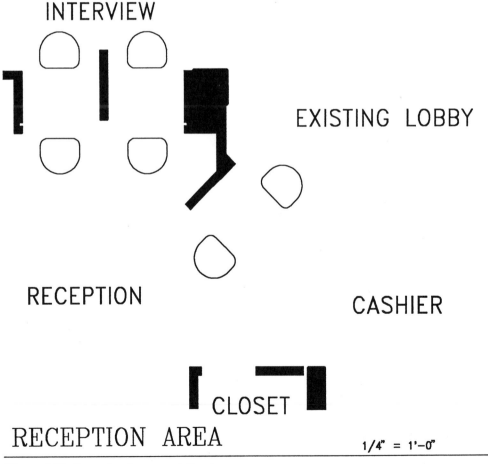

Figure 6.35 Example of a presentation layer.

INTERVIEW

EXISTING LOBBY

RECEPTION

CASHIER

CLOSET

RECEPTION AREA

1/4" = 1'-0"

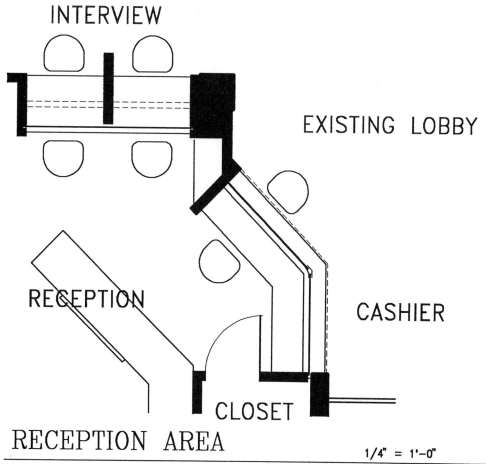

Figure 6.36 *Example of a presentation drawing using architectural base and unique presentation layer.*

The Code Overlay

Another use for the architectural base plan is the *code plan*. Using the architectural base plan, prepare an overlay showing agency-specific data such as occupancy type and count, fire-rated walls, area separations, exiting, and so on. This plan can be photographically reduced by 50 percent, composited with the reduced overlay, and blown back on a new drawing film along with floors 2, 3, and 4. In this way, the code plans can be drawn by utilizing the architectural base plan and photoreprographics. (Remember to make all lettering twice the size on the code overlays so that they are correct size when reduced.)

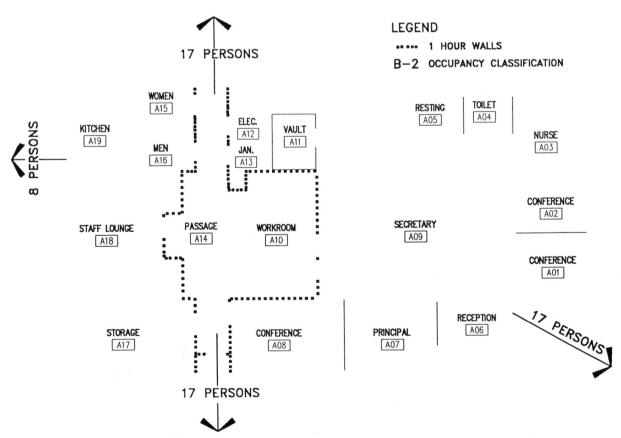

Figure 6.37 *Example of a code layer.*

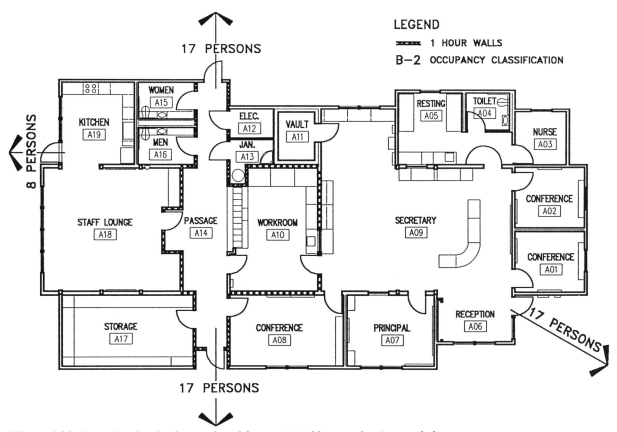

Figure 6.38 *Example of code plans reduced from original base and unique code layers.*

PLANNING FOR OVERLAY PRODUCTION

The uses a drawing base can be put to are limited only by imagination. The technology is there to produce a large variety of products from one drawing. The most important requirement of overlay production to keep in mind is *planning*. Drawings are no longer single sheets, but multiple layers. With this complexity comes a need for greater control of the production process.

Preplanning the Project

Preplanning a project is very important. The best person to help you do this is your reprographics manager. This person understands the media and can help you to avoid costly production errors.

I know of a project which was drawn by using overlay production. The project manager did not understand the compositing process and ordered architectural overlays to be combined with each engineering overlay. The resulting (full-value) blackline prints made no sense at all, and the whole order had to be reprinted. All of this happened on a very large, out-of-state project with a short bid period. The mistake cut a whole week off the bid process and caused the printing costs to double.

Attend Training Seminars

Also, select a person on each project team, preferably the production leader, to establish and monitor the overlay process for each project. Have this person attend training seminars on overlay drafting and be the liaison with the reprographics house.

Use the Reprographics Manager to Assist with Planning

If there are any questions regarding the materials, systems, or methods of reprographics, contact your reprographics manager for help.

When a project is started, meet with your reprographics manager to discuss the project schedule and your printing needs. Schedule printing activities well ahead of the time you expect to need them. Often, conflicts arise from assuming the reprographics can be done at any time. Print shops are busy places, and they need advance notice to guarantee your schedule. Also, overlay reprographics takes longer than straight diazo copying. This extra time must be planned for or the print order will be late.

Also, talk about the drawing media you propose to use (see Chap. 4). For best results, use polyester drawing film and pen and ink for line work. Use plastic lead pencils for lettering. No graphite. If drawings are to be screened, do not use dot pattern appliqués. They will either fade out completely or become solid when run through a second screen.

Plan Layering with All Team Members

The success of overlay production also depends on the cooperation of consulting engineers. If they do not agree to produce their drawings using overlays and architect's backgrounds, most of the advantage is lost. All of the potential uses for in-house layering still remain, and are good reasons to proceed, but the cost of reproduction might not be considered a part of the reimbursable expenses unless it is "needed" to composite consultant drawings too. So use consultants who will agree to produce their drawings using your base sheets and who understand the requirements of overlay production.

Project Start-Up Coordination Meeting

At project start-up, discuss the overlay process with key team members including consulting engineers. Explain the overlay system and the way it is expected to be managed. This discussion should include the following items:

Distribution and retrieval of background slicks should be scheduled depending on the size and complexity of the project. Never allow a consultant to keep an old slick. They will use it, and this results in mistakes. Always collect the old slicks and file them for record until the project is complete.

Determine which drawings will be produced by overlay. Floor plans are a natural, but how about reflected ceiling plans? Some architects produce ceiling overlays which are drawn over the base floor plans. This allows them to use the wall lines already drawn as ceiling borders. Others like to draw a ceiling base drawing with all new line work. In either case, the consultant's drawings for duct layouts, lighting, etc. are produced on their own unique sheet. The engineers need to accept the architect's method of producing ceilings, so do not proceed until this issue is resolved.

Drafting media must be compatible (see Chap. 4). Consultants must use pen and ink and not graphite lead for drawing so that camera images are compatible.

Publish the project schedule for printing at the close of schematics, design development, and contract documents. Allow plenty of time for printing and checking the documents before they are delivered.

Cartooning for Overlay Drafting

When planning the project, cartoon the set in a manner similar to the procedures outlined earlier in this chapter. Since some drawings will be produced using overlays, have some cartoon sheets printed on tracing stock. These can then be sketched on, like a miniature overlay is, to graphically show everyone what the set, is supposed to look like. Once the cartoons are done, prepare a drawing index, but this time do it on the print-order matrix.

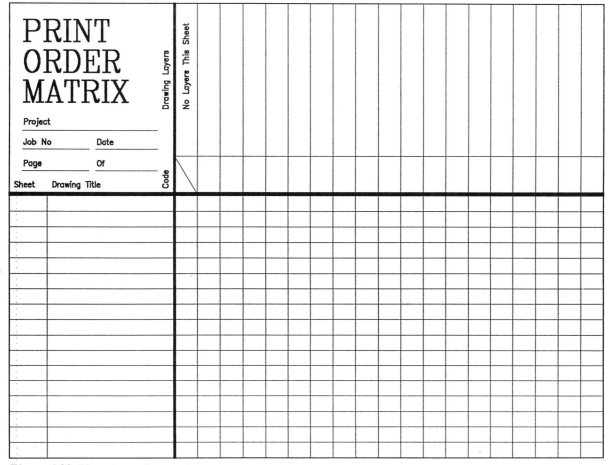

Figure 6.39 *The print-order matrix.*

Completing the Print-Order Matrix

List all drawings on *the print-order matrix,* whether they are produced in layers or on single sheets. (See Chap. 7 for the recommended drawing order.) Enter sheet number and drawing title in the column provided on the left side of the form. In the top row, enter every drawing layer, base sheets first, followed by overlay(s) for each drawing title in the left column. List the layers in the same order as the drawing-sheet list with one exception: Do not repeat entries for architectural base plans used for plumbing or mechanical or electrical drawings. Each drawing layer is only listed one time.

PRINT ORDER MATRIX	Drawing Layers →	No Layers This Sheet	SITE SURVEY	SITE BASE	LEVEL – 1 BASE PLAN	LEVEL – 1 PLAN NOTES	LEVEL – 2 BASE PLAN	LEVEL – 2 PLAN NOTES	ROOF BASE PLAN	ROOF PLAN NOTES	ARCHITECTURAL SCHEDULES	LEVEL – 1 PLUMBING	LEVEL – 2 PLUMBING	LEVEL – 1 HVAC	LEVEL – 2 HVAC	ROOF HVAC				
Sheet / Drawing Title (Code)																				
A1	SITE PLAN																			
A2	LEVEL 1 PLAN																			
A3	LEVEL 2 PLAN																			
A4	ROOF PLAN																			
A5	BUILDING ELEVATIONS																			
A6	BUILDING SECTIONS																			
A7	WALL SECTIONS																			
A8	EXTERIOR DETAILS																			
A9	WINDOW DETAILS																			
A10	STAIRS & ELEVATORS																			
A11	SCHEDULES																			
A12	INTERIOR DETAILS																			
M1	LEVEL 1 PLUMBING PLAN																			
M2	LEVEL 2 PLUMBING PLAN																			
M3	LEVEL 1 HVAC PLAN																			
M4	LEVEL 2 HVAC PLAN																			
M5	ROOF HVAC PLAN																			

Figure 6.40 The print-order matrix.

Use of Film Negatives

Some drawings or drawing layers may already exist as an 8½-inch-by-11-inch negative. These often come from previous projects or from the Library of Standards. If this is the case, enter negatives to be used as layers in the far right portion of the form.

PRINT ORDER MATRIX		Drawing Layers	No Layers This Sheet	SITE SURVEY	SITE BASE	LEVEL – 1 BASE PLAN	LEVEL – 1 PLAN NOTES	LEVEL – 2 BASE PLAN	LEVEL – 2 PLAN NOTES	ROOF BASE PLAN	ROOF PLAN NOTES	ARCHITECTURAL SCHEDULES	LEVEL – 1 PLUMBING	LEVEL – 2 PLUMBING	LEVEL – 1 HVAC	LEVEL – 2 HVAC	ROOF HVAC	FINISH SCHEDULE MASTER	OPENING SCHEDULE MASTER	STD TITLE & BORDER
Project																				
Job No	Date																			
Page	Of	Code																		
Sheet	**Drawing Title**																			
A1	SITE PLAN																			
A2	LEVEL 1 PLAN																			
A3	LEVEL 2 PLAN																			
A4	ROOF PLAN																			
A5	BUILDING ELEVATIONS																			
A6	BUILDING SECTIONS																			
A7	WALL SECTIONS																			
A8	EXTERIOR DETAILS																			
A9	WINDOW DETAILS																			
A10	STAIRS & ELEVATORS																			
A11	SCHEDULES																			
A12	INTERIOR DETAILS																			
M1	LEVEL 1 PLUMBING PLAN																			
M2	LEVEL 2 PLUMBING PLAN																			
M3	LEVEL 1 HVAC PLAN																			
M4	LEVEL 2 HVAC PLAN																			
M5	ROOF HVAC PLAN																			

Figure 6.41 Show negatives in the far right columns.

The Drawing/Layer Code

Under each drawing description is a space for the *drawing code assignment.* This code has two parts. The first letter designates whether the drawing is a base or an overlay. Base sheets are designated "B." Overlays are identified by a letter reflecting the discipline that it originated with, i.e., "A" for architectural and "C" for civil. The second part is a number representing the specific overlay from that discipline. Simply number consecutively from the first overlay to the last in a consultant's series. Other, more complex systems may be developed that help identify floor or wing of a building.

Sheet	Drawing Title	No Layers This Sheet	SITE SURVEY (SS)	SITE BASE (SB)	LEVEL – 1 BASE PLAN (B–1)	LEVEL – 1 PLAN NOTES (A–1)	LEVEL – 2 BASE PLAN (B–2)	LEVEL – 2 PLAN NOTES (A–2)	ROOF BASE PLAN (B–3)	ROOF PLAN NOTES (A–3)	ARCHITECTURAL SCHEDULES (AS–1)	LEVEL – 1 PLUMBING (P–1)	LEVEL – 2 PLUMBING (P–2)	LEVEL – 1 HVAC (M–1)	LEVEL – 2 HVAC (M–2)	ROOF HVAC (M–3)	FINISH SCHEDULE MASTER (NEG–3)	OPENING SCHEDULE MASTER (NEG–2)	STD TITLE & BORDER (NEG–1)
A1	SITE PLAN																		
A2	LEVEL 1 PLAN																		
A3	LEVEL 2 PLAN																		
A4	ROOF PLAN																		
A5	BUILDING ELEVATIONS																		
A6	BUILDING SECTIONS																		
A7	WALL SECTIONS																		
A8	EXTERIOR DETAILS																		
A9	WINDOW DETAILS																		
A10	STAIRS & ELEVATORS																		
A11	SCHEDULES																		
A12	INTERIOR DETAILS																		
M1	LEVEL 1 PLUMBING PLAN																		
M2	LEVEL 2 PLUMBING PLAN																		
M3	LEVEL 1 HVAC PLAN																		
M4	LEVEL 2 HVAC PLAN																		
M5	ROOF HVAC PLAN																		

Figure 6.42 Fill in the drawing-layer code.

The Match Box

The drawing/layer code is also used to identify each drawing layer on a special form located in the binding edge of the drawing sheet. This small form is called the *match box.* Its purpose is to provide space to identify the base sheet drawing and all overlays that belong to it. This match box could serve as a key for placing every drawing sheet in its proper combination. For example, everyone has had a stack of film drawings slide off the drawing board only to be scattered all over the floor. With overlay drafting there is only one title block and it appears on the base sheet. Without another method for identifying every base and layer with a unique number, that pile of drawings on the floor might just as well stay there. But the match-box numbers make it possible to reassemble the drawings. The match box appears on the most common sheet. Generally this is the architectural base drawing. Never place more than one match box per drawing title sheet. They will not align. Within the match box is a column of prompts followed by six additional columns for recording the base and overlays needed to make a drawing.

MATCH BOX		BASE	OVERLAY	OVERLAY	OVERLAY	OVERLAY	OVERLAY
	SCREEN						
	DRAWING LAYER						
	THIS LAYER						

Figure 6.43 The match box.

Enter Base Layers on Match Box

Using the print-order matrix as a guide and starting from the bottom of the match box, the first prompt is for THIS LAYER. The first layer(s) to number in a drawing set are the base layers. Looking again at the Urban Banking Company as an example, the first-floor plan base layer will have a "B1" in the first column next to THIS LAYER. The second-floor base layer will have a "B2" in this same position, and so on, until the roof plan which will have a "B5" entered.

MATCH BOX		BASE	OVERLAY	OVERLAY	OVERLAY	OVERLAY	OVERLAY
	SCREEN						
	DRAWING LAYER						
	THIS LAYER	B1					

Figure 6.44 Match box with base plan indicated.

Enter Unique Layers on Match Box

Go back to the first-floor plan architectural base. Place it on the drawing board and align it with the pin bar. Next, place the architectural unique layer for this plan over the base plan and align it with the pin bar. Reading through the top layer, the match box is visible on the layer below along with the entry "B1." Enter an "A1" (taken from the print-order matrix) in the first column marked OVERLAYS and in the same row marked THIS LAYER. In the next row marked DRAWING LAYERS enter an "A1" directly above the other "A1."

In the column above the "B1" on the base plan, enter another "B1" on the overlay drawing. These entries describe the layers that make up the drawing. Whenever both rows are "matched" (hence the name "match box"), the drawing is properly assembled. This is also why the unique layer has an entry in both rows.

MATCH BOX		BASE	OVERLAY	OVERLAY	OVERLAY	OVERLAY	OVERLAY
	SCREEN						
	DRAWING LAYER	B1	A1				
	THIS LAYER	B1	A1				

Figure 6.45 Enter the unique layer and base reference.

Enter Consultant Layer Codes to Match Box

Remove the "A1" overlay and align the consultant's plan overlays one at a time. Enter a "P1" for plumbing layer, "M1" for mechanical layer, and so on until all first-floor layers have been identified. Remember, each unique layer is identified in the same position on the match box. They are never printed together, so there is no conflict. Also enter the base-layer number (B1), as before, to complete the "match." These two entries are all that is needed to find the proper layers that make up a drawing. Regular diazo copying can be done with no more information.

Mechanical Plan

MATCH BOX		BASE	OVERLAY	OVERLAY	OVERLAY	OVERLAY	OVERLAY
	SCREEN						
	DRAWING LAYER	B1	M1				
	THIS LAYER	B1	M1				

Plumbing Plan

MATCH BOX		BASE	OVERLAY	OVERLAY	OVERLAY	OVERLAY	OVERLAY
	SCREEN						
	DRAWING LAYER	B1	P1				
	THIS LAYER	B1	P1				

Architectural Plan

MATCH BOX		BASE	OVERLAY	OVERLAY	OVERLAY	OVERLAY	OVERLAY
	SCREEN						
	DRAWING LAYER	B1	A1				
	THIS LAYER	B1	A1				

Figure 6.46 Match boxes for each consultant.

Enter Requirements for Screen Backgrounds

Screens are generally used to make the architectural base layer read weaker than consultant overlays, thereby making the main subject read better. When screening is desired, this requirement must also be added to the match-box data. In the column used for each drawing layer is space for SPECIAL SCREENS. Enter the basic screen percent, i.e., 50 or 30, whichever is required. Make sure that your reprographics manager understands exactly what this means. If no screen is needed, enter a "_" or an "X" to indicate that the screen was considered and was not wanted.

MATCH BOX		BASE	OVERLAY	OVERLAY	OVERLAY	OVERLAY	OVERLAY
	SCREEN	30%	X				
	DRAWING LAYER	B1	A1				
	THIS LAYER	B1	A1				

Figure 6.47 Enter screen value for backgrounds.

COMPOSITE OPTIONS

Progress printing from overlay drawings can be copied in two basic ways: *direct copy* and *intermediate copy.*

Direct copy can be done in the office by exposing light-sensitive paper to the drawing images on a flat-bed printer (see Chap. 5) and developing it in the diazo printer.

Assume, for example, that the drawing to be copied is the first-floor plan of the Urban Banking Company. Follow the steps outlined below:

     \#\#\#   Place the architectural base plan (B1 in the match box) face-down on the frosted-glass surface of the flat-bed printer.

     \#\#\#   Place the architectural unique layer (A1) face-down on top of the base and register them with plastic "buttons" through at least two of the holes at the top edge of the sheet. This order is important. The layer that is closest to the copy medium will be the most clear. It is the notes, dimensions, etc. that should read the best, so this unique layer is closest to the print paper.

     \#\#\#   Place a sheet of copy paper face-down over the originals and expose according to the exposure schedule.

     \#\#\#   Remove the copy paper and develop it in the diazo machine.

This process can be repeated as often as necessary, but if more than three copies are needed, there is a better way: Make an intermediate.

Making Intermediates from Drawing Layers

Intermediates are reproducible copies of an original drawing or layers of original drawings from which opaque copies are made. When multiple copies are needed of the first-floor plan, make an intermediate by doing the following:

    ###   Place the originals in the same order but face-up.

    ###   Cover with intermediate reproduction stock, face-down.

    ###   Register with the plastic buttons.

    ###   Make the exposure.

    ###   Develop in the diazo copier.

The intermediate will be reverse-reading, which produces better copies when run through the diazo machine as an original.

Evaluate Compositing Quality and Price

For final printing, the direct-copy method would take far too long and would not really produce the best copies. Intermediates are both fast and superior in quality. The paper sepia intermediate is okay for progress printing, but again, it does not yield the best results. Diazo slicks or clear-polyester-film intermediates produced in sepia tone yield very good copies. The next best intermediate is the photographic wash-off or fixed-image on polyester film. (See Chap. 5 for complete review of materials and uses.) Finally, for the best quality in printing architectural drawings, use the offset-printing process.

The offset-printing process makes the clearest possible copies and they can be reproduced in color. This combination makes them desirable for very complex work, but there is a price to pay. The offset process is very expensive and not within most budgets for reprographics.

The Standard Drawing Matrix

The overlay-printing process is substantially more complex than straight diazo printing. The print-order matrix is the form most used to express all the requirements for combining and then copying drawing layers. The form is used in the following three ways:

As the standard drawing matrix

As a diazo print order

As a photo composite order

As the standard drawing matrix, the field of the form is completed by placing a dot, "X," check, or other mark in the boxes that associate a drawing title with its base, overlays, and negatives. It is that simple. By reading the form, anyone can tell what it takes to create a drawing. It becomes graphically obvious which drawings have two overlays, or no overlays at all. This form becomes a good tool in planning drawing production and for recording drawing progress. For example, a budget of time can be assigned to each element on the form and monitored.

PRINT ORDER MATRIX

Project ___ Job No ___ Date ___ Page ___ Of ___

Sheet	Drawing Title	Code	No Layers This Sheet	SITE SURVEY SS	SITE BASE SB	LEVEL – 1 BASE PLAN B-1	LEVEL – 1 PLAN NOTES A-1	LEVEL – 2 BASE PLAN B-2	LEVEL – 2 PLAN NOTES A-2	ROOF BASE PLAN B-3	ROOF PLAN NOTES A-3	ARCHITECTURAL SCHEDULES AS-1	LEVEL – 1 PLUMBING P-1	LEVEL – 2 PLUMBING P-2	LEVEL – 1 HVAC M-1	LEVEL – 2 HVAC M-2	ROOF HVAC M-3	FINISH SCHEDULE MASTER NEG-3	OPENING SCHEDULE MASTER NEG-2	STD TITLE & BORDER NEG-1
A1	SITE PLAN			•	•															•
A2	LEVEL 1 PLAN					•	•													•
A3	LEVEL 2 PLAN							•	•											•
A4	ROOF PLAN									•	•									•
A5	BUILDING ELEVATIONS		•																	
A6	BUILDING SECTIONS		•																	
A7	WALL SECTIONS		•																	
A8	EXTERIOR DETAILS		•																	
A9	WINDOW DETAILS		•																	
A10	STAIRS & ELEVATORS		•																	
A11	SCHEDULES											•						•	•	•
A12	INTERIOR DETAILS		•																	
M1	LEVEL 1 PLUMBING PLAN					•							•							
M2	LEVEL 2 PLUMBING PLAN							•						•						
M3	LEVEL 1 HVAC PLAN					•									•					
M4	LEVEL 2 HVAC PLAN							•								•				
M5	ROOF HVAC PLAN									•							•			

Figure 6.48 The print-order form used as a standard drawing matrix.

The Diazo Print Order

As a diazo print order, the field is completed by entering a number representing the first, second, or third layer away from the reprographic medium. Remember, the layer closest to the print medium will reproduce the clearest. This is designated "number 1" on the print-order matrix. The remaining layer(s) are designated 2 and 3 in order of expected clarity. When completed, this form becomes the reprographic order form.

PRINT ORDER MATRIX

Project ___
Job No ___ Date ___
Page ___ Of ___

Sheet	Drawing Title	Code	No Layers This Sheet	SITE SURVEY (SS)	SITE BASE (SB)	LEVEL – 1 BASE PLAN (B-1)	LEVEL – 1 PLAN NOTES (A-1)	LEVEL – 2 BASE PLAN (B-2)	LEVEL – 2 PLAN NOTES (A-2)	ROOF BASE PLAN (B-3)	ROOF PLAN NOTES (A-3)	ARCHITECTURAL SCHEDULES (AS-1)	LEVEL – 1 PLUMBING (P-1)	LEVEL – 2 PLUMBING (P-2)	LEVEL – 1 HVAC (M-1)	LEVEL – 2 HVAC (M-2)	ROOF HVAC (M-3)	FINISH SCHEDULE MASTER (NEG-3)	OPENING SCHEDULE MASTER (NEG-2)	STD TITLE & BORDER (NEG-1)
A1	SITE PLAN			2	1															3
A2	LEVEL 1 PLAN					2	1													3
A3	LEVEL 2 PLAN							2	1											3
A4	ROOF PLAN									2	1									3
A5	BUILDING ELEVATIONS		1																	
A6	BUILDING SECTIONS		1																	
A7	WALL SECTIONS		1																	
A8	EXTERIOR DETAILS		1																	
A9	WINDOW DETAILS		1																	
A10	STAIRS & ELEVATORS		1																	
A11	SCHEDULES											1						2	3	4
A12	INTERIOR DETAILS		1																	
M1	LEVEL 1 PLUMBING PLAN					2							1							
M2	LEVEL 2 PLUMBING PLAN							2						1						
M3	LEVEL 1 HVAC PLAN					2									1					
M4	LEVEL 2 HVAC PLAN							2								1				
M5	ROOF HVAC PLAN									2							1			

Figure 6.49 The print-order form used as a diazo print order.

The Photo Composite Order

As a photo composite order, the field is completed with a dot, "X," check, or other mark to associate a drawing title with its base and overlays. Screening is indicated by stating the screen percent (copy from match box) for each layer. When completed, this becomes the reprographic order form.

PRINT ORDER MATRIX

Project _____

Job No _____ Date _____

Page _____ Of _____

Sheet	Drawing Title	Code	No Layers This Sheet	SITE SURVEY (SS)	SITE BASE (SB)	LEVEL–1 BASE PLAN (B-1)	LEVEL–1 PLAN NOTES (A-1)	LEVEL–2 BASE PLAN (B-2)	LEVEL–2 PLAN NOTES (A-2)	ROOF BASE PLAN (B-3)	ROOF PLAN NOTES (A-3)	ARCHITECTURAL SCHEDULES (AS-1)	LEVEL–1 PLUMBING (P-1)	LEVEL–2 PLUMBING (P-2)	LEVEL–1 HVAC (M-1)	LEVEL–2 HVAC (M-2)	ROOF HVAC (M-3)	FINISH SCHEDULE MASTER (NEG-3)	OPENING SCHEDULE MASTER (NEG-2)	STD TITLE & BORDER (NEG-1)
A1	SITE PLAN			50%	✓															100%
A2	LEVEL 1 PLAN					100%	✓													100%
A3	LEVEL 2 PLAN							100%	✓											100%
A4	ROOF PLAN									100%	✓									100%
A5	BUILDING ELEVATIONS		✓	.																
A6	BUILDING SECTIONS		✓																	
A7	WALL SECTIONS		✓																	
A8	EXTERIOR DETAILS		✓																	
A9	WINDOW DETAILS		✓																	
A10	STAIRS & ELEVATORS		✓																	
A11	SCHEDULES											✓						50%	50%	100%
A12	INTERIOR DETAILS		✓																	
M1	LEVEL 1 PLUMBING PLAN					30%							✓							
M2	LEVEL 2 PLUMBING PLAN							30%						✓						
M3	LEVEL 1 HVAC PLAN					30%									✓					
M4	LEVEL 2 HVAC PLAN							30%								✓				
M5	ROOF HVAC PLAN									30%							✓			

Figure 6.50 The print-order form used as a photo composite order.

Recap

The process of separating drawings into reusable base elements and unique overlays allows the design team to produce their documents at a reduced cost when compared to conventional drafting, while increasing the degree of coordination between the documents. The process is complex and needs to be monitored by someone who understands how it works. To assist the architect, a meeting should be held between all design-team members to review the goals of the overlay system and determine how it will be executed on a day-to-day basis. Use these decisions to cartoon a miniature set of drawings and to fill in the print-order matrix. Contact the reprographics manager early in the process to help plan the production and reproduction processes as well as scheduling print sessions. By coordinating the overlay process, it can be a real asset to the production process of any architectural practice.

CAD PRODUCTION

A series of events in the 1980s has brought on a revolution in architectural production as important as systems production and overlay drafting. These events center on the introduction of the affordable personal computer and other related hardware and software, allowing most architectural offices to begin drafting by machine.

The first computer-aided drafting systems were installed on mainframe computers and used with remote work stations. They were very expensive, and only a few of the largest firms could afford them. Today, the hardware has been reduced in size to fit on a desktop or even on a laptop, with enough power to produce complex floor-plan drawings for large buildings. Personal computers carry enough memory to run very sophisticated CAD programs and store the drawing files. Hardcopy drawings must be plotted so that they can be used by someone other than the person preparing the drawing. This is accomplished on either a pen plotter for large drawings, or a desktop printer for small-size drawings and quick-check copies.

ESTABLISH A CAD SYSTEM

The CAD Committee

One of the first steps in using CAD is to appoint a CAD manager. This person should be placed directly under the quality-control manager and should work with the quality-control manager in committee with two or three top CAD users to establish CAD production standards. As before, with systems production, an advisory member to this decision-making body is needed: the reprographics manager. This person can help you to avoid costly situations and find ways out when they are unavoidable.

Level of Commitment

This CAD committee should meet to determine the level of commitment the firm will give to computer-aided drafting. This is a big step for most architectural and engineering firms, and must be considered carefully. There are a number of elements to review before making such a decision.

> ### Without looking at systems, what kind of costs can be invested on a process that could fail? There is no guarantee that comes with CAD stating that you will increase profits or reduce production time. An entry-level system with one CAD station, pen plotter, and laser printer will likely cost about the same as one production architect's annual salary.

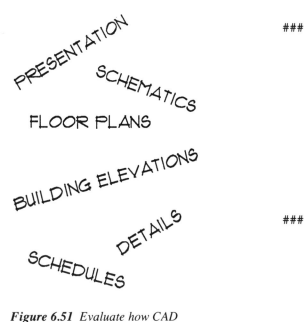

Figure 6.51 *Evaluate how CAD will be implemented into office production.*

Need. Why is CAD being considered? If it is only to keep up with the competition, there is likely not a sufficient mind-set to make it work. There must be a desire to learn complex routines on sophisticated equipment to perform most of the same tasks you now perform quite well with ink on drawing film. With a real desire to learn CAD and an enjoyment of computer operations, it can be made to work better than hand drawing. Without this desire, the joy would probably turn to a feeling of being burdened with extra equipment.

How will CAD be used? It would be a mistake to attempt a whole set of drawings on CAD without going through a learning course on partial sets first. Start with floor plans for a relatively simple building, and master that first. Then proceed with building elevations, sections, and details. CAD can also be used to receive files from word-processor and database programs so that schedules can be produced by CAD. Determine the number of new features that you can successfully learn and master on any one project.

Who will learn CAD? It takes time to learn computer-aided drafting, and many project managers and even lead production staff often feel that they do not have the time to become proficient with CAD. They may be right. There are three directions to go. Build a CAD production pool where trained (schooled) CAD production "drafters" produce architectural drawings at the direction of "architects" who do only sketch-level drawings. Or, require all architects to learn CAD and give each a work station complete with CAD, word processor, spreadsheet, and database. Managers will use CAD less and the other programs more than production staff, but they will all become computer-literate. A combination of CAD pool and CAD architect is likely a third option.

How will CAD be integrated with consultants? The CAD committee should hold meetings with each consulting discipline to see how they are using CAD and to determine how they can interface with the architect. Items to discuss include CAD programs, hardware problems, layer systems, and graphic standards. These questions should be kept basic and general in nature. Later, when a project is being set up, other discussions will be needed to work out the fine points.

CAD HARDWARE SYSTEM ACQUISITION

Invite product vendors to show off their systems. Have a list ready of the things you expect CAD to do for you. If possible, have someone you trust, who knows CAD, sit in and help ask questions. Every system has its good points and some not so good. They are difficult to evaluate without some experience.

Basic CAD Station

Most CAD systems now run on microcomputers with a single monitor and mouse for control. Some of the more powerful set-ups include a second monitor and digitizer tablet with arrays of commands and functions. Some CAD systems also work on laptop computers with mouse control and have excellent graphic quality. These work well when conferencing with clients on their turf and when you want to use the power of CAD to evaluate options.

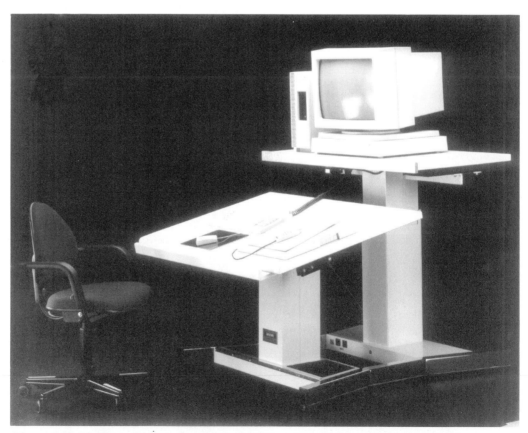

Figure 6.52 *The basic CAD station*. (The basic CAD station courtesy of Mayline Hamilton.)

Printers/Plotters

With the exception of laptops, it is not feasible to bring the computer with you to show someone the drawings. They must be plotted in large-sheet format or printed in 8½"-by-11" format. Check the options for full-size plotters. Some use pens and others are electrostatic for faster plotting in blackline only. Pen plotters also come in a table model for smaller drawings. Tabletop printers use either dot-matrix technology for draft-quality prints or laser technology for final-drawing quality. Color printers are also available.

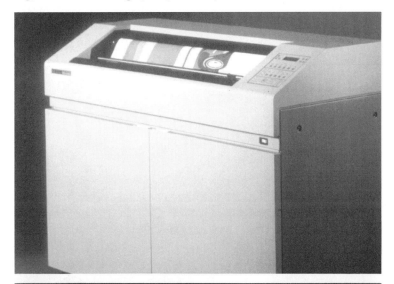

Figure 6.53 *Printers and plotters.* (Printers and plotters courtesy of Hewlett-Packard Company.)

Computer Links

Unless you plan to have only one CAD station, consider the method for sharing your data files. For example, one architect is working dimensions on the floor plan and another architect wants to work on the reflected ceiling plan which uses the same architectural base drawing. This can be done by using computer links between work stations. One station is set up as the main file holder and the others borrow information much like from a library. The difference is that others can still "see" the files that someone else is working on. This process can be taken one step further. Consider the use of a modem, so outside consultants can access your link and share data too. (More on this when we discuss layering systems.)

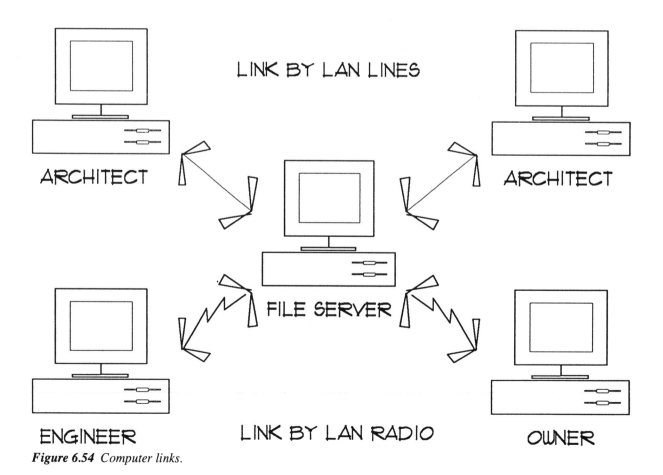

Figure 6.54 Computer links.

Continuing Education

The CAD committee can be expanded, with time, to five persons. The new members should include persons using other software so that their involvement in CAD is represented. One member from this committee should attend local meetings of CAD users to help keep the firm on the cutting edge of CAD production. The committee should also subscribe to user magazines. These often give helpful hints on how to solve problems. This committee should establish the budget recommendations to management for all computer usage. Budget items should include hardware and software acquisitions, plotter supplies, training, periodicals, memberships, and the like. Consider sending CAD persons to special outside courses in advanced CAD usage.

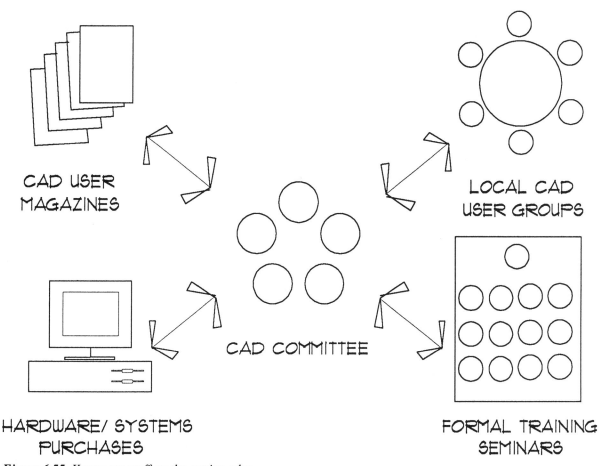

CAD USER
MAGAZINES

LOCAL CAD
USER GROUPS

CAD COMMITTEE

HARDWARE/ SYSTEMS
PURCHASES

FORMAL TRAINING
SEMINARS

Figure 6.55 *Keep your staff on the cutting edge.*

CAD GRAPHICS

The graphics used in CAD production should be the same as those used in manual production. Following is a review of line-work lettering, symbols, poche, and the standard drawing sheet.

Line Work

CAD production standards include every element of drawing from line work to layering, so persons using the machine have the "tools" at their fingertips, just like with manual drafting. The first element to standardize is line weight and configuration. There are four basic line weights for manual drafting. These same line weights must be assigned to the CAD system so that the operator can draw with them. This then allows for hand drafting in ink on CAD plots. The system must also have various line types for easy recall and use including the following: continuous, dashed, hidden, and center-lines. CAD programs allow the operator to program line weights and configurations for easy recall and use. The method for doing this varies by CAD system and can be found in the CAD training manual.

LINE	PEN SIZE	PEN NO.	APPLICAICATIONS
————	0.25	3X0	USED FOR DIMENSION LINES, NOTE LEADERS, CEILING GRIDS, ITEMS IN ELEVATION
————	0.35	0	OUTLINE ITEMS ABOVE THE FLOOR IN PLAN OUTLINE ITEMS AWAY FROM WALLS IN ELEVATION FURNITURE, EQUIPMENT, FIXTURES
————	0.50	1	WALLS AND DOORS IN PLAN WALLS IN REFLECTED CEILING PLAN OUTLINE IN BUILDING SECTION VIEW
————	0.70	3	BUILDING OUTLINE IN SITE PLAN OUTLINE IN WALL SECTION AND DETAIL CLOUDS AROUND REVISIONS

Figure 6.56 Line weight and configurations.

Lettering

Before lettering can begin, a number of discussions must be made about the font to be used. By definition, a *font* is a collection of characters and symbols which are described by a common type-face, weight, style, and size. Type faces include such standards as Helvetica and Times Roman. The font weight is usually listed as fine, regular, and bold. Style refers to normal, italics, or sloped. Lettering size is entered in one of two ways. Points are the common measure for font sizes used in graphic production. In this system, 72-point lettering is one-inch tall, 36-point lettering is ½-inch tall, and 18-point lettering is ¼-inch tall. In CAD production, lettering is more often described as being "X"-inches tall.

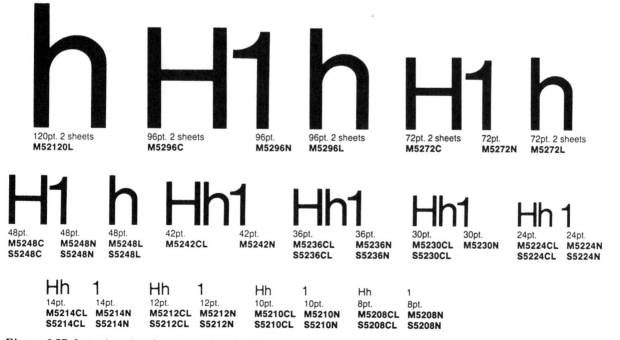

Figure 6.57 *Lettering size chart.* (Lettering size chart courtesy of Chart Pak, Inc.)

Lettering size is a product of the scale at which a drawing will be plotted. For example, if a floor plan is drawn at ⅛-inch scale, minimum lettering of ⅛-inch-high will be entered as 12-inches-high (12 drawing inches = ⅛ inches actual). Room names which are ³/₁₆-inches in real life will be entered as 16-inches-high. Drawing titles which are ¼-inch-high will be entered as 24-inches-high. A similar progression works for drawings of any scale.

Because of this characteristic of CAD, a drawing produced for ⅛-inch scale plotting will still have ⅛-inch-high lettering when plotted at either ¹/₁₆-inch scale or ¼-inch scale. As drawings are scaled down, the lettering becomes crowded and often overlaps. As drawings are scaled up, lettering can look relatively small, and the drawing appears unfinished.

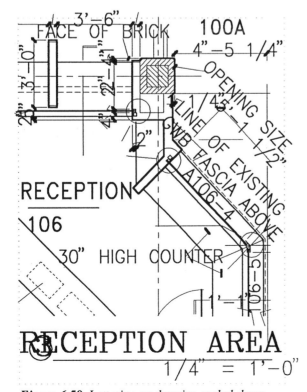

Figure 6.58 Lettering on drawing scaled-down.

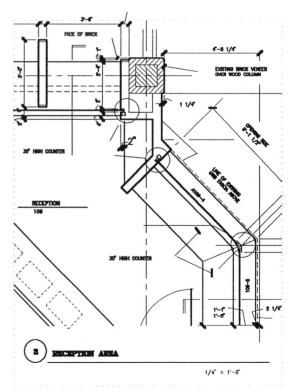

Figure 6.59 Lettering on drawing scaled-up.

Many CAD systems have provisions for telling the system what scale the drawing is to be plotted at so lettering is automatically sized up or down if the drawing is enlarged or reduced. This can have its downside too. Now as a plan is reduced, the lettering is too small to be clearly read and when the drawing is enlarged, all lettering looks like titles.

The answer to the lettering-scale problem is to know how drawings will be used, what the CAD system will do with the lettering, and to plan around it. This might mean entering repeat room titles on separate layers for each scale that the drawing will be plotted at. When recording the layers used to produce a specific drawing, include only the correct plan-text layers.

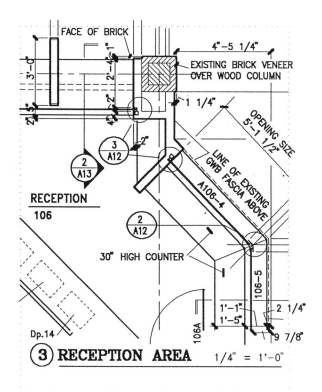

Figure 6.60 Lettering on drawing scaled-down.

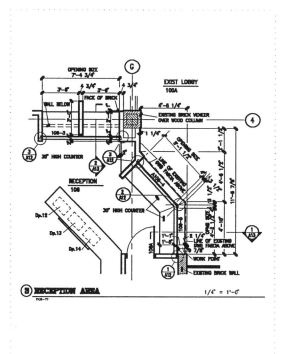

Figure 6.61 Lettering on drawing scaled-up.

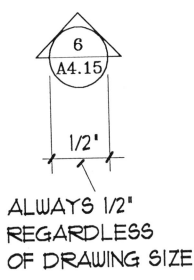

ALWAYS 1/2"
REGARDLESS
OF DRAWING SIZE

Figure 6.62 *Program reference symbols to drawing scale.*

Symbols

It is easy for manual drafters to use a template and quickly draw the symbols they need. In CAD production, if the symbol doesn't exist in memory, it can take a long time to draw it from scratch. Computer-aided drafting can perfectly reproduce graphic symbols every time they are needed, and because they are computer-generated, they can be made to ask the operator questions about symbol contents. For example, the section indicator can be formatted to prompt the operator for both section number and drawing-sheet number. Before this can be done, however, a standard group of graphic symbols must be established. Produce a list of CAD standard symbols for all staff to use.

Produce CAD symbols that match the graphic symbols in Chap. 1. Program each referencing symbol to prompt the user for drawing identification and sheet number. Also, program the systems to draw the symbol at the appropriate scale. For example, a plumbing fixture is drawn at real scale, but detail references are programmed to be ½-inch in diameter regardless of drawing scale. The same applies to the lettering inside the detail and other reference indicators. Once the graphic symbols and lettering sizes are created, the next element to establish is poché patterns.

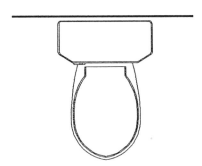

ALWAYS 1:1 TO
SCALE WITH
DRAWINGS

Figure 6.63 *Program plumbing symbols to real scale.*

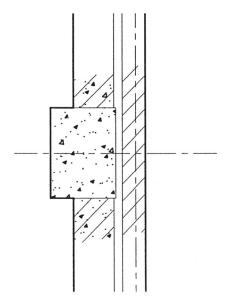

Figure 6.64 *Partial plan with poché.*

Poché

Poché by computer is time-consuming (relatively speaking) and requires a great deal of computer memory. For this reason, many architects prefer to omit poché from their drawings. This, however, removes a very important element from the graphic language and must be replaced by notes or other methods of communication. Rather than omitting poché, it is better to find a way to manage it. Poché with straight lines is the fastest to draw and uses the least amount of computer memory. Curved lines are the slowest to draw and require the most computer memory. Assuming that poché patterns described in Chap. 1 are to be used for both manual drafting and computer drafting, then the way to reduce memory clog and to speed up the poché process is to limit the amount of the poché pattern used on a drawing. Consider poché only at ends of a material or wall, and at intersections with other materials. This could reduce the amount of poché to 50 percent and more compared to what it would take to poché an entire drawing. Where full poché is desired, consider making the poché another drawing file separate from the basic drawing file. The two files can then be recombined for plotting. The important issue here is to use the same poché for both manual and CAD production. Your specific CAD system and computer hardware might handle this problem with better or worse results. Work with them to develop a method for using the poché patterns selected as standard.

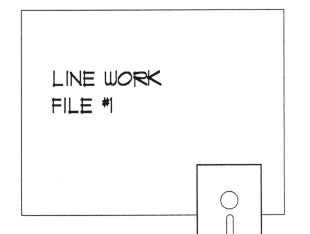
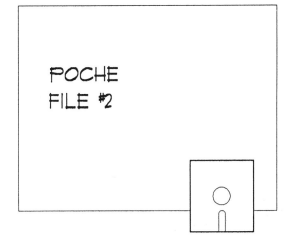

Figure 6.65 *Consider two drawing files for heavy poché drawings.*

Drawing Sheet and Title Block

The next graphic to create is the *standard drawing sheet.* Again, see Chap. 1 for examples with various sheet sizes. Add the standard drawing matrix shown earlier in this chapter. All these elements are used at real scale. That means that a 24-inch-by-36-inch drawing sheet used with an 1/8-inch scale drawing will be drawn at real scale sized 192-feet-by-288-feet. A 30-inch-by-42-inch drawing sheet will be 240-feet-by-336-feet. This is calculated as eight times the sheet size in inches equals the drawing size in feet. A 1/4-inch scale drawing sheet will be 96-feet-by-144-feet for 24-inches-by-36-inches and 120-feet-by-168-feet for a 30-inch-by-42-inch.

Start by drawing the title block at the desired actual size i.e., 24 × 36 or 30 × 42. Then scale it up to drawing scale following the procedure just mentioned. When completed, be sure that all lines are drawn with the proper pen and line type and save for future use.

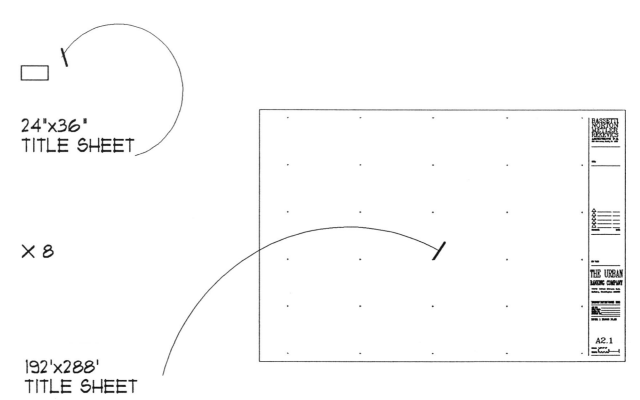

24"x36"
TITLE SHEET

× 8

192'x288'
TITLE SHEET

Figure 6.66 *Draw title block at actual size and scale up to drawing scale.*

Managing a CAD Production System

CAD has taken the concepts of overlay production and amplified them in computer format. Where in manual overlay production only three layers are recommended for best reprographic clarity. In CAD production, hundreds of layers of information can be made and recomposited internally in any order or combination without affecting the final quality of production.

So far, the tools established will allow an operator to draw almost anything using standard line work on a standard drawing sheet, including standard drawing symbols and references. Next, the CAD layer system must be established. This will allow a degree of flexibility in architectural production never before available. The process begins, as did overlay production, with a meeting between the architect and the consulting engineers. With 100 or more layers to work with, there is a need to affectively manage the system.

CAD Layering

Layering of information, as in overlay drafting, enhances the flexibility of the information by allowing it to be used in a greater number of ways. Some of the benefits of layering are the following:

- The ability to select information by subject and to combine it with other information to create new drawings

- To coordinate the work of various design disciplines through shared information

- To share designated information with other production people, all at the same time

- To use repeat information as in a multistory building

Layer Identification

The method used to identify layers is either by numbering or by naming. Some CAD systems require one or the other, and this should become an issue when purchasing CAD from a vendor. The numbering method is generally more restrictive when it is imposed by the CAD system. For systems that allow both numbering and naming, the number of possible layers can be over 1000.

Layer Sets

When creating a layer system, consider how it will be used. First of all, there are many disciplines that will all need to draw and share information. A good place to start layer identification is with *layer sets* assigned to each discipline. Identify each layer set by using the same letter(s) referenced in Chap. 7 to preface their sheet numbers.

A	Architectural
C	Civil
CA	Cabinetwork
E	Electrical
EQ	Equipment
I	Interiors
FS	Food Service
FP	Fire Protection
GR	Graphics
L	Landscape
M	Mechanical (Air Handling)
MC	Mechanization
P	Plumbing
S	Structural
VT	Vertical Transportation

Layers, like EQ and CA, are used by vendors who now quite regularly have their own "add-ons" to some CAD programs. They can either produce their own drawings, or the architect can do this, using their software. In either case, a layer set is designated for this use. Later, this information can be used to make up its own drawing sheet, or be added to the architectural plans.

View

Layer information is next divided into *Views* which indicate which of the following a drawing is:

P Plan
E Elevation
S Section
D Detail
X Schedule

This division will separate the information so that it is apparent which layer is used for what. For example, drawing elements in detail are not the same as their smaller-scale plan or section counterparts. They, therefore, belong on a different layer to avoid using the wrong poche or line weights between the three views.

LAYER SET

A Architectural
C Civil
CA Cabinet Work
E Electrical
EQ Equipment
I Interiors
FS Food Service
FP Fire Protection
GR Graphics
L Landscape
M Mechanical
MC Mechanization
P Plumbing
S Structural
VT Vertical Transport

VIEW

P Plan
E Elevation
S Section
D Detail
X Schedule

CATEGORY

DRAWING RELATED LAYERS
TB Title & Border
GD Grid
ML Match Line
DM Dimensions
DE Demolition Notes
AR Area Calculations
CD Code Notes & Graphics
RE Reference Symbols
HA Hatch Patterns
PL Plotter Information
NT Notes

BUILDING RELATED LAYERS
WA Wall
DR Doors & Relites
WW Windows
CL Ceiling

Figure 6.67 Layering chart.

Category

The third element of layer identification is the *category.* This identifies the unique element being drawn as a wall, door, counter-top, or room title. There are individual lists for each layer set. For example, light fixtures are only found in layer set "E." In this way, the light-fixture symbol drawn by the electrical engineer is shared by the architect and the mechanical engineer on their drawings. Each list is divided into two parts, one for drawing-related layers, the other for building-related layers.

Drawing-related items include:

TB	Title and Border
GD	Grid
ML	Match Line
DM	Dimensions
DE	Demolition Notes
AR	Area Calculations
CD	Code Notes and Graphics
RE	Reference Symbols
HA	Hatch Patterns
PL	Plotter Information
NT	Notes

Building-related layers include:

WA	Wall
DR	Door and Relites
WW	Window, Storefronts, Curtain Walls
CL	Ceiling

Recap

So far, we have discussed the first three elements of a layer system:

###	Layer Set
###	View
###	Category

Sample Floor Plan Layers

Within this system, an architectural floor plan would have the following layers:

A,P,WA	Architectural, Plan, Walls
A,P,DR	Architectural, Plan, Doors, and Relites
A,P,WW	Architectural, Plan, Windows
A,P,GD	Architectural, Plan, Grids
A,P,DM	Architectural, Plan, Dimensions
A,P,RE	Architectural, Plan, Reference Symbols
A,P,HA	Architectural, Plan, Hatch Patterns
A,-,TB	Architectural, ——, Title, and Border
A,-,NT	Architectural, ——, Notes

The Room Finish Schedule information should be placed in the following layer:

A,X,FN   Architectural, Schedule, Finishes.

The Door and Opening Schedule would be drawn in:

A,X,DR   Architectural, Schedule, Doors, and Relites.

Continue with this direction when making up other drawing layers. Create a master list from which to run all jobs.

Drawing Layer Make-up Matrix

As in overlay drafting (see previous section in this chapter), a matrix of drawing titles and layers is prepared. This *Drawing Make-up Matrix* lists every drawing required in the project along the left side, and drawing layer, along the top. Use the Drawing Make-up Matrix to plan drawing layering and to provide a guide for compositing drawing sheets.

The example below shows these drawing sheets, but you might find it best to keep drawings of a kind together, i.e., floor plans and building elevations.

Sheet	Drawing Title	—TB	APBL	APGR	APPV	APDM	APNT	APRE	APHA	APWA	APDR	APWW	APCA	AEWA	AEDR	AEWW	AEDM	AENT	AERE
A1	SITE PLAN	●	●	●	●	●	●	●											
A2	LEVEL 1 PLAN	●				●	●	●	●	●	●	●	●						
A3	LEVEL 2 PLAN	●				●	●	●	●	●	●	●	●						
A4	ROOF PLAN	●												●	●	●	●	●	●
A5	BUILDING ELEVATIONS	●												●	●	●	●	●	●
A6	BUILDING SECTIONS																		
A7	WALL SECTIONS																		
A8	EXTERIOR DETAILS																		
A9	WINDOW DETAILS																		
A10	STAIRS & ELEVATORS																		
A11	SCHEDULES																		
A12	INTERIOR DETAILS																		
M1	LEVEL 1 PLUMBING PLAN																		
M2	LEVEL 2 PLUMBING PLAN																		
M3	LEVEL 1 HVAC PLAN																		
M4	LEVEL 2 HVAC PLAN																		
M5	ROOF HVAC PLAN																		

Column headers (Drawing Layers):
- —TB : NO CAD LAYER THIS SHEET / SHEET TITLE & BORDER
- APBL : ARCH PLAN BLDG LINE
- APGR : ARCH PLAN GRADING
- APPV : ARCH PLAN PAVEMENT
- APDM : ARCH PLAN DIMENSIONS
- APNT : ARCH PLAN NOTES
- APRE : ARCH PLAN REF SYMBOLS
- APHA : ARCH PLAN HATCH PATTERNS
- APWA : ARCH PLAN WALLS
- APDR : ARDH PLAN DOORS & RELITES
- APWW : ARCH PLAN WINDOWS
- APCA : ARCH PLAN CABINET WORK
- AEWA : ARCH ELEV WALLS
- AEDR : ARCH ELEV DOORS
- AEWW : ARCH ELEV WINDOWS
- AEDM : ARCH ELEV DIMENSIONS
- AENT : ARCH ELEV NOTES
- AERE : ARCH ELEV REF SYMBOLS

LAYER MAKE—UP MATRIX

Project _____
Job No _____ Date _____
Page _____ Of _____
Drawing Layers / Code

Figure 6.68 Drawing Layers Make-up Matrix.

Figure 6.69 *A typical pen plotter.*
(A typical pen plotter courtesy of Hewlett-Packard Company.)

PLOTTING

Plotting is generally accomplished on machines known as *pen plotters.* These machines carry a carousel which holds between six and eight drawing pens. They produce drawings of various sizes depending on the size of the machine. Some plotters are small, tabletop devices used for 8½-inch-by-11-inch plots. Large plotters are console units, and are capable of producing drawings to 36-inches-by-48-inches. These plotters work on a variety of paper and polyester film products and use a variety of pen types and colors. (See Chap. 4 for descriptions of production products.) The large variety of products allows the user to plot in color or black and white, for reproduction, or a one-of-a-kind drawing.

Plotting can also be done on electrostatic plotters. Their advantage is speed. They are up to 200 times faster than pen plotters. This can be very important when a large number of plots are required in relatively short time. In this way, drawings that take two or more hours to plot may only take minutes on an electrostatic plotter. The output is black-on-white and is not erasable. It is, therefore, not the medium of choice for drawings that might require hand revisions later. Electrostatic plots are great for progress printing when they are used only to make copies and to store as a record. One more thing: If the electrostatic plotter is not adjusted properly, the drawings it produces may be slightly out-of-scale. This makes these plots unusable for any overlay work. Pen plotters may be slow, at least compared to electrostatic plots, but they are still the more versatile of the two.

Drawings can also be obtained by printing on a dot-matrix or laser printer. Dot-matrix printers offer the largest variety of drawing sizes because they accept a greater variety of paper sizes than laser printers. These include: 8½-inch-by-11-inch, 8½-inch-by-14-inch, and 11-inch-by-17-inch. Most laser printers can use 8½-inch-by-11-inch and 8½-inch-by-14-inch only. Of the two printers, the laser printer can be made to emulate full-sheet plotters, producing line work that varies in weight as if drawn by differing pen sizes. The final image is easy to read and excellent for reproduction processes. Dot printers cannot vary line weight and often produce a drawing that is difficult to read because of the resolution of the machine. For quick prints of details and other drawings, this may not be a problem and is more economical than laser printing.

Figure 6.70 *A typical laser printer.* (A typical laser printer courtesy of Hewlett-Packard Company.)

CAP

Computers are used for more than just producing drawings. They can be used at the very start of a project for any of the following:

Accounting

Fact finding

Scheduling

Correspondence

Project planning

There are many application programs available for use in *Computer-Aided Production* (CAP) which can assist the architect in every aspect of project development. The first problem is deciding which software packages are best for each application. The CAD committee should be expanded to address these questions:

What budget is available for computer-aided production?

Who will use computers?

How will these systems and programs be used?

Budget for Four Basic Programs

The budget for CAP is not so large as it is for CAD. Most of the hardware is the same, but the software is unique to the need. For example, spreadsheet applications usually do not perform word processing, and database managers do not work as project managers. These programs, however, can be made to work together and with CAD. In this way, information from one program can be shared with other programs. For example, information can be typed into a word processor where editing is easy, and copied into CAD for a Room Finish Schedule where editing is not so easy. For the tops in performance, it is best to consider purchasing a top-quality program from each of the function categories, being sure that files are transferable between them. These functional categories are:

Word Processor

Project Manager

Spreadsheet

Database Manager

Who Will Use CAP?

The discussion about who will use computers and for what, needs a great deal of thought. Accounting and word processing are two obvious uses, so the people involved with each of these need a computer. The project-management applications are best used by managers who have their own equipment. Database applications can be used by all staff architects to gather and schedule information, so computers should be available. Consider a computer station where everyone can come to share the resources.

How Will CAP Be Used?

CAP can help the most in dealing with the massive quantities of information that make up a project. Begin with the Room Data Form and Equipment Data Form. Enter information into a computerized version of these two forms. From them, production schedules can be created, including the following:

Door and Opening Schedule

Room Finish Schedule

Equipment Schedule

These schedules are actually using the same information entered into the database. This information is only entered once. It is only the format for printing the information that changes, i.e., final schedule form. In this way, data is entered at the time of programming and kept active and accurate until document publication and building construction. This continual link helps to guarantee the results. Data entered at programming *is* the data used for schedules, including revisions. Consult your local software vendors on programs that can perform this function.

If every project manager and at least his or her leading technical architect has a computer capable of performing database entry and manipulation, word processing, spreadsheet analysis and project management, the process of information handling can become easier and more accurate.

CONCLUSION

The systems and tools used in architectural contract document production are many and varied. The economics of running a modern professional practice demand that staff and management become familiar with those items discussed here, and learn to use them to their advantage. To run an office without them is to invite financial and legal problems. Using the systems and methods discussed here will help any architect avoid mistakes, produce a project more quickly, and potentially leave more time for solving design problems.

DOCUMENT IDENTIFICATION AND ORDER

7

INTRODUCTION

The most common drawing-sheet numbering system has architectural drawings begin with an "A," structural drawings begin with an "S," mechanical drawings begin with an "M," and electrical drawings begin with an "E." This brief system is quite adequate for many projects, but not for those in which the project becomes complex or when the drawings are issued in phases. Every project has its own individual requirements for drawing numbering. Because every proj-ect is different, it will often become necessary to modify the basic system to fit the individual project.

FUNCTIONS OF A DRAWING NUMBER

Two of the primary functions of the drawing number are to prevent preliminary drawings from being used as contract documents and to identify drawings by major discipline. The system most often used for this purpose is one which has a two-part preface, beginning with one or two letters, used to indicate the phase of work. The code for these letters is taken from the list below. No preface is used for the basic contract document set, as this is the base document.

PHASE PREFACE CHART

MARK	PHASE OF WORK
PR	Programming
MP	Master Plan
S	Schematic
D	Design Development
*	Contract Documents
R	Revision

*No preface is used for contract document phase.

The second part of the preface consists of one or two letters used to designate the design discipline responsible for the drawings. This letter(s) is taken from the list below and is entered directly after the phase letter(s). This list is shown alphabetically, and does not represent drawing order.

DESIGN DISCIPLINE CHART

MARK	DESIGN DISCIPLINE
A	Architectural
C	Civil
CA	Cabinetwork
E	Electrical
EQ	Equipment
FS	Food Service
FP	Fire Protection
GR	Graphics
I	Interiors
L	Landscape
M	Mechanical (Air Handling)
P	Plumbing
S	Structural
VT	Vertical Transportation

Example Sheet Numbers

The basic drawing numbering system is completed by adding the next consecutive sheet numbers. The following examples illustrate drawing sheet numbers for various phases of work and design discipline.

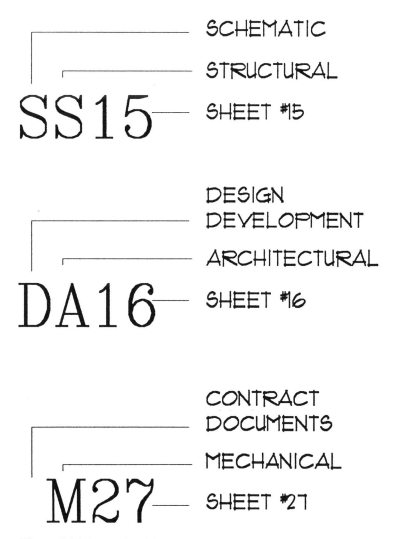

Figure 7.01 Example of drawing numbers.

DRAWING ORDER

The drawings of each primary design discipline are placed into a set in a very traditional order. This order is:

C Civil
A Architectural
S Structural
M Mechanical
E Electrical

See Chap. 8 for a description of the contents of each group of drawings.

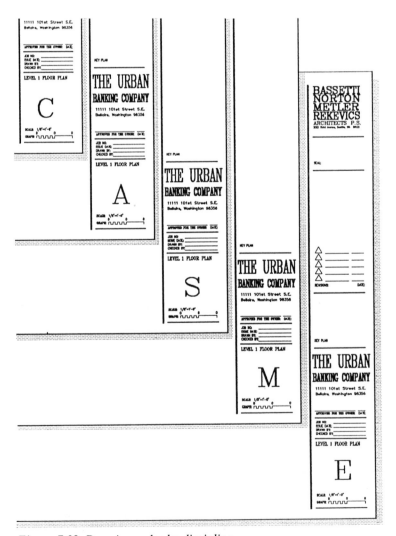

Figure 7.02 *Drawing order by discipline.*

Drawing Order for Complex Projects

As the complexity of the building increases there is often the need for more design consultants with their own package of drawings. Locate these drawings into the basic drawing order. For example, if landscape drawings are produced, they usually follow the civil set. If fire-protection drawings are needed, they usually follow the mechanical set.

A set of drawings for a complex institutional project like a hospital might contain civil, landscape, architectural, interiors, graphics, mechanical, food-service, fire-protection, and electrical drawings. The drawing set should be arranged in the order shown below. Use the preface letter(s) noted.

MARK	DESIGN DISCIPLINE
*	Cover Sheet
C	Civil
L	Landscape
A	Architectural
FS	Food Service
I	Interiors
G	Graphics
S	Structural
EQ	Equipment
VT	Vertical Transportation
M	Mechanical
P	Plumbing
FP	Fire Protection
E	Electrical

*No page number

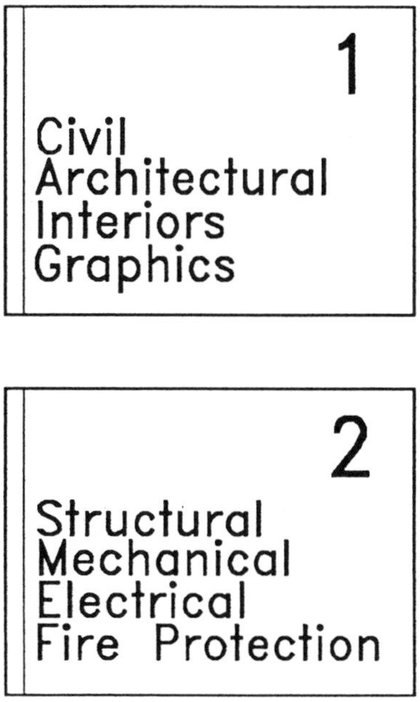

Binding the Drawings

Often the drawings are too numerous to bind into one set. When this happens, consider dividing them into two or more volumes. A good place to divide the sets is between architectural and structural. Each set will have its own cover sheet noting volume 1 and volume 2. A side benefit to this two-volume order is that it puts the structural set at the top of the pile, and this, along with the civil drawings in volume 1, are the two sets that the contractor uses the most at the beginning of a job.

1
Civil
Architectural
Interiors
Graphics

2
Structural
Mechanical
Electrical
Fire Protection

Figure 7.03 *Divide drawings into two sets.*

CONSULTANTS' DRAWING ORDER

Within the drawing disciplines, information is shown in specific locations which are generally accepted. For example, the civil package starts with the site survey followed by site demolition, building location, grading, and site utilities. The number of drawings depends on the complexity of the site. If the project is being phased, locate these drawings after the survey.

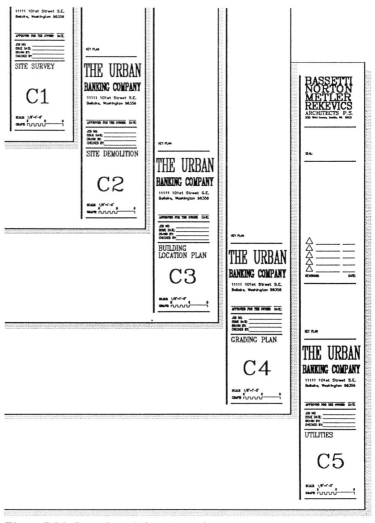

Figure 7.04 *Consultants' drawing order.*

Structural Drawing Order

Structural drawings begin with an overall foundation plan followed by floor and roof framing plans, building sections, and wall sections. Details start with the Footing Schedule, followed by the Column Schedule and the Beam Schedule. The detail packages are grouped by structural material, i.e., concrete, steel, and wood.

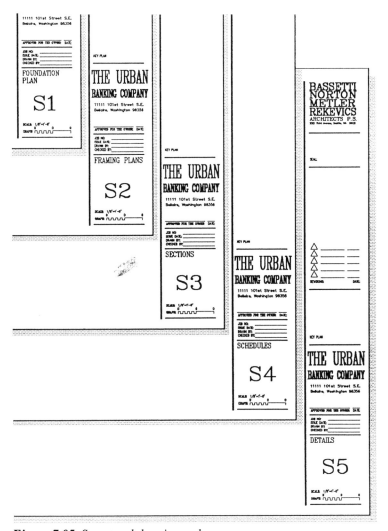

Figure 7.05 *Structural drawing order.*

Mechanical Drawing Order

The mechanical set starts with building, heating, ventilation, and air-conditioning plans, followed by schedules and details. Next are plumbing drawings and riser diagrams. Mechanical drawings usually do not include fire-protection sprinkler drawings. There is a liability issue here, and the construction industry and insurance companies have developed a practice whereby sprinkler systems are usually designed by the contractor who provides installation. The mechanical drawings will show water service to the sprinkler shut-off valve, and the sprinkler design contractor will design the rest of the system. For more on this see Chap. 10, "Quality-Control," for fire-protection systems.

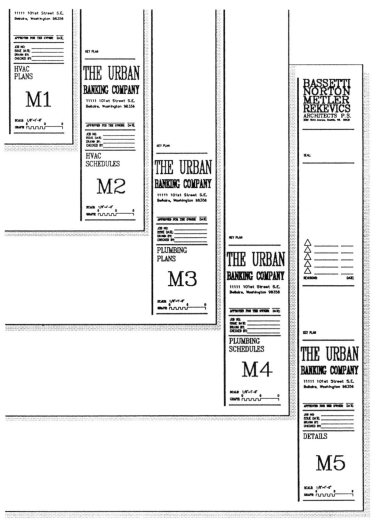

Figure 7.06 *Mechanical drawing order.*

Electrical Drawing Order

The electrical set starts with the building lighting plans followed by fixture schedules. Next are power-distribution plans, panel schedules, and details. Communications drawings are either prepared by the electrical engineer or a communications vendor and follow the power drawings when prepared by the engineer.

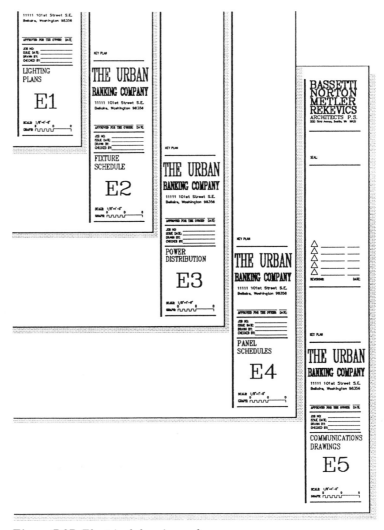

Figure 7.07 *Electrical drawing order*

ARCHITECTURAL DRAWING ORDER

The architectural drawing set is divided into three general categories which represent the three basic areas of construction. These are:

Site
Building
Interiors

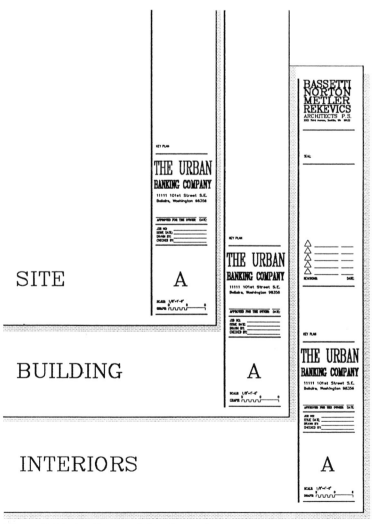

Figure 7.08 *Architectural drawing order.*

Site Drawings

The drawing set begins with a site plan. For small projects with simple site development, all site improvements can be shown on a single drawing. For more complex projects, the site plan may be divided into several drawings beginning with the basic architectural plan followed by detail plans for stairs, ramps, planters, and the like.

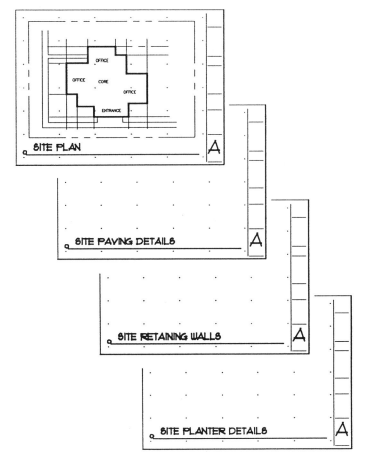

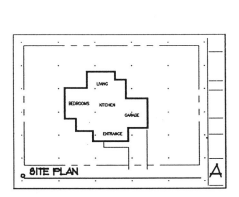

SIMPLE PROJECT

COMPLEX COMMERCIAL PROJECT

Figure 7.09 Site development drawings.

Building Drawings

The building is defined by a series of drawings. These are floor and roof plans, building elevations, building sections, wall sections, and building details. This package contains the basic information to build the building shell.

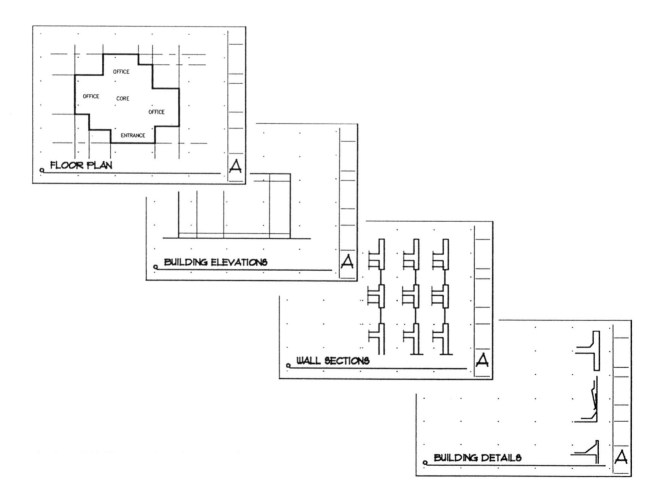

Figure 7.10 *Building development drawings.*

Interior Drawing

The interior package includes schedules for finishes, door and opening, and partition, color and material plans and elevations, detail plans and elevations, reflected ceiling plan, and interior details.

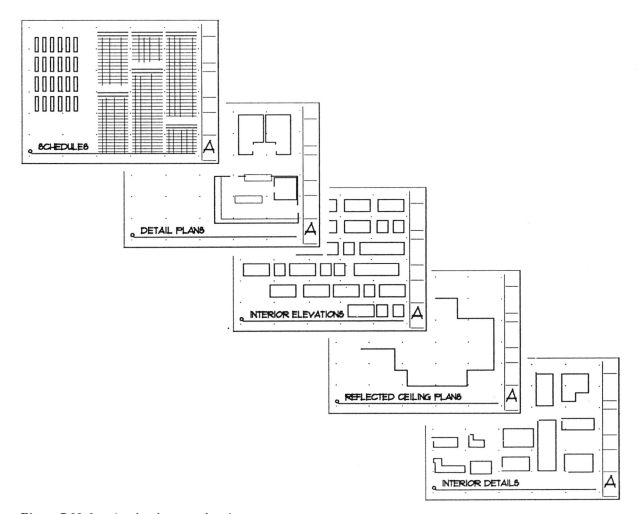

Figure 7.11 *Interior development drawings.*

Drawing Index

The following example represents a set of drawings for a relatively simple building:

MARK	DRAWING NAME
A0	Drawing Index, Symbols, Abbreviations, Design-Team Credits, General Notes
A1	Site Plan and Details
A2	Floor Plan and Roof Plan
A3	Building Elevations and Sections
A4	Wall Sections and Details
A5	Finish Schedule, Door Schedule, Details
A6	Detail Plans and Interior Details
A7	Reflected Ceiling Plan

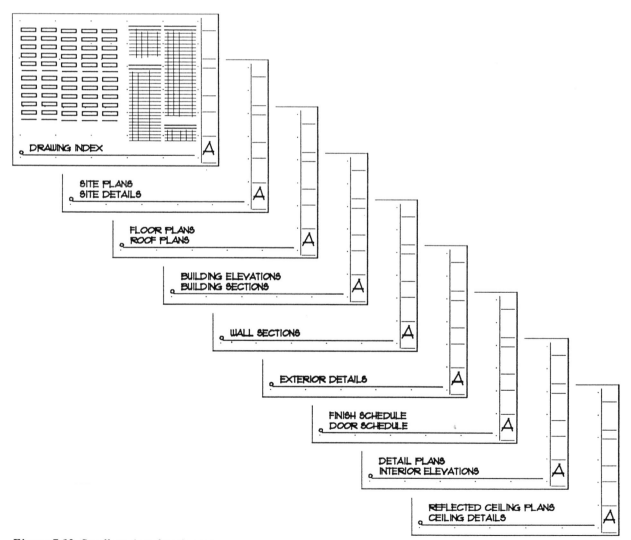

Figure 7.12 Small-project drawing set.

Follow the Three Basic Elements of a Building Set

Notice the group and arrangement of the three basic drawing types.

      ###   Sheet A1 defines the site.

      ###   A2, A3, and A4 define the building.

      ###   A5, A6, and A7 define the interiors.

Try to avoid cross-overs between groups. For example, if the floor-plan sheet has room, use it for shell details, not for interior details or schedules. If the interior schedule sheet has room left, don't use it for the Window Schedule.

You will likely have some arguments with your peers regarding this philosophy, and there are as many variations in drawing order as there are building designs. This particular system, like any system, is not the *only* way to order the set. The important message here is, by establishing a higher logic to the drawing order, the contractor and owner will find the set easier to read.

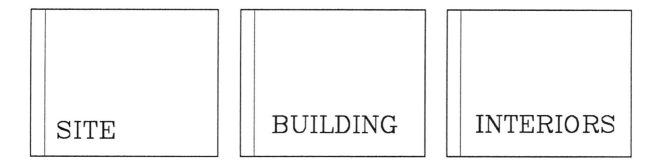

Figure 7.13 Follow the basic drawing order.

Problems with Consecutive Numbering

On very large projects, a consecutive drawing-numbering system breaks down as soon as an additional drawing sheet is added and must be inserted within the set. It is possible to number the new drawings by adding a decimal suffix such as A37.2, A37.3, etc. These numbers do provide a unique address for the drawings within the set, but they violate the standard-sheet-numbering order for the project. Ostensively, this appears to be a good system. However, there are cases where it can cause confusion. For example, if you were to loose drawing number A37.3 in the series listed above, it would not be apparent that a drawing was missing because it is not consecutively numbered. A person refiling the drawings would simply go on from A37.2 to drawing number A38 without realizing that A37.3 was missing. This same mistake could be made at the print shop where drawings are copied and assembled. Should the drawing be inadvertently misplaced, sheet number A37.3 could be omitted from the set without the print operator realizing the mistake.

System Numbering for Large Projects

A system has been invented which solves the problem of consecutive numbering by dividing the drawings into chapters. The chapters listed in the chart below follow the basic sheet layout described for small projects.

CHAPTER	TITLE
A0	General
A1	Site Conditions
A2	Floor Plans, Roof Plan
A3	Building Elevations and Sections
A4	Wall Sections, Panels
A5	Building Details, Schedules
A6	Detail Plans
A7	Interior Elevations
A8	Reflected Ceiling Plans
A9	Interior Details, Schedules

This chapter system also supports the three-main-drawing-grouping philosophy. Chapter A1 defines the site, Chaps. A2, A3, A4, and A5 define the building, and Chaps. A6, A7, A8, and A9 define the interiors.

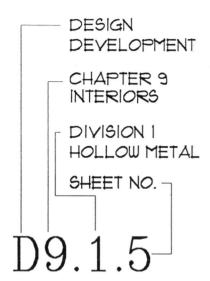

Figure 7.14 A design-development architectural sheet number in chapter 2.

Expand on the Chapter System

When the chapters do not adequately divide the set, use subdivisions within chapters. For example, a large jail will have a large hollow-metal package. This is best shown in a subdivision of Chap. 9 as 9.1—Hollow Metal. Then the remaining interior details must be placed in 9.2—Interior Details. The same applies to exterior details. A restoration of an ornate terra-cotta building may require a subdivision of Chap. 5, for example, 5.1—Terra-Cotta Restoration. Then the remaining exteriors can be grouped into 5.2—Exterior Details.

Stick with the System Once Started

If the chapter system is selected, even if only one chapter is very complicated, use all of the chapter numbers as outlined. Do this even if some chapters have only one drawing. The reason is simple. Within the office, you can always find wall sections in Chap. 4, and this commonality is important in reducing production time and assisting in quality control. If, by chance, a whole chapter is not used, note this in the drawing index. For example, Chap. 6—not used.

Within the chapter system, each drawing sheet is numbered consecutively. The example below shows a design development drawing in Chap. 2. Once you know the system, it is easy to identify the basic group by the chapter number.

Special Conditions

Most of the problems encountered in using the chapter system of ordering drawings come in two specific chapters. These are Chapters 5 (Building Details and Schedules) and 9 (Interior Details and Schedules). As noted above, some projects have at least one area of work which requires a breakdown of one of these chapters. In reality, large projects will have numerous subdivisions within Chapters 5 and 9. This can become confusing, so another system exists to deal with the contents of these chapters.

The Detail-Book Method of Publishing Details and Schedules

This alternate system of schedule and detail development is called the *Detail Book*. Within this system, the project has large-size drawing sheets and a book consisting of schedules, notes, and details. The large drawings use Chaps. A0 through A4, and A6 through A8. Chapters A5 and A9 are noted as follows.

CHAPTER	TITLE
A5	Building Details and Schedules (see Detail Book)
A9	Interior Details and Schedules (see Detail Book)

Alternative Systems Exist

The detail book can be ordered in several ways. Numerous authors and organizations including ConDoc™ and GUIDELINES™ have proposed systems for numbering and ordering construction details. ConDoc™ uses the MASTERSPEC™ numbering system which the American Institute of Architects recommends as a specification outline. GUIDELINES™ uses an adaptation of the specification-outline system promoted by the Construction Specification Institute (CSI).

These two systems are similar in that they take a specification order and attempt to use it to order construction details. These specification systems work well for their intended purpose: organizing building elements for the purpose of writing specifications. But they are not well-suited for detail numbering.

Specifications Do Not Represent Assemblies

The reason the specification systems do not work well for detail numbering is simple. Specifications describe individual components of a building. A detail is a composite of these components. Attempting to number a detail by selecting a primary component is difficult to do. For example, take a roof-edge detail which shows roofing, blocking, stucco wall, sealant, building framing, and the roof gravel stop. This detail has many primary components, and the drawing could be filed in any one of at least four specification categories. As a result, it could be easily lost, and it could take far more time to find than if it were filed in another way. As a composite of building materials, a detail needs a system of identification which allows it to be filed as an assembly. For this purpose, a 10-chapter detail-book index system is proposed here instead of a specification ordering system.

The 10-Chapter Detail-Book System

The 10-chapter detail-book system focuses on details as composites of building materials which follow the drawing order established for large-sheet drawings. That order is as follows:

Site
Building
Interiors

The detail book, like all other drawings, is begun with a chapter for general items. This chapter is followed by nine other chapters which are divided up as shown in the chart below.

CHAPTER	TITLE
0	General
1	Site Conditions
2	Exterior Conditions
3	Moisture Protection
4	Windows, Storefronts, Louvers
5	Doors, Relites, Folding Partitions
6	Stairs and Railings
7	Interior Conditions
8	Specialties
9	Casework and Millwork
10	Equipment

Notice that Chap. 1 defines the site, Chaps. 2, 3, and 4 define the building, and Chapters 5, 6, 7, 8, 9, and 10 define interior conditions. This is a very logical and easy-to-use system for arranging construction details into book form.

Expansion of the 10-Chapter System

Within the 10 chapters there are subdivisions where any detail can be filed. These subcategories are permanent addresses used within an office detail system and are never changed. For example, door details are always found in Chap. 5, and interior finishes are always found in Chap. 7. If you had a project which did not have a door, the contents of Chap. 5 would be listed as "not used," but the number would not be eliminated. In this same project, as in every project, the interior finishes will still be in Chap. 7. The following is a full and complete listing of all detail book chapters and subheadings.

MASTER DETAIL INDEX

0 GENERAL

0-1	General Architectural Drawing Index
0-1S	General Structural Drawing Index
0-1M	General Mechanical Drawing Index
0-1E	General Electrical Drawing Index
0-2	General Notes
0-3	Graphic Standards
0-4	Abbreviations
0-5	Legal Descriptions
0-6	Project Identification
0-7	Temporary Facilities

1 SITEWORK

1-1	Site Drainage (CB, YD)
1-2	Pavement and Walks
1-3	Site Furniture and Improvements
1-4	Landscaping

2 EXTERIOR CONDITIONS

2-1	Concrete (CIP, Precast)
2-2	Masonry (CMU, Brick, Stone)
2-3	Misc. Metal Louver Schedules
2-4	Wood, Stucco, Plastic
2-5	EQJ/Exterior
2-6	Prefab Panel Systems (GFRC, Brick)

3 MOISTURE PROTECTION

3-1	Membrane Roofing
3-2	Roof Accessories (Hatches)
3-3	EQJ/Roof
3-4	Membranes (Sheet, Fluid)
3-5	Shingles (Wood, Tile, Composition)
3-6	Metal Roofing

4 WINDOWS

4-0   Schedules
4-1   Metal Sash
4-2   Storefront, Curtain wall
4-3   Wood Sash
4-4   Miscellaneous (Atrium)

5 DOORS, RELITES, AND OTHER OPENINGS

5-0   Schedules
5-1   Hollow Metal
5-2   Wood
5-3   Specialties (Roll-up, Overhead)

6 STAIRS AND RAILINGS

6-0   Schedules
6-1   Concrete Stairs
6-2   Metal Stairs
6-3   Wood Stairs
6-4   Ladders, Catwalks
6-5   Railings

7 INTERIOR CONDITIONS

7-0   Schedules
7-1   Floor and Base
7-2   Walls, Partitions, Furring
7-3   Ceilings, Soffits
7-4   EQJ/Interior

8 SPECIALTIES

8-1   Toilet Partitions and Accessories
8-2   Conveying Systems (Elevators)
8-3   Operable Walls, Folding Doors
8-4   Misc. Specialties
      Curtain Tracks
      Drapery Tracks
      IV Tracks
      Chalk and Tack Boards
      Dock Facilities
      Chutes
      Access Panels

9 CASEWORK AND MILLWORK

9-0	Schedules
9-1	PLAM Casework
9-2	Metal Casework
9-3	Hardwood Casework
9-4	Demountable Casework
9-5	Millwork

10 EQUIPMENT

10-0	Schedules
10-1	Fixed Equipment

When the details are organized under this system, a Table of Contents is included in Chap. 0. This table lists every drawing in the detail book. It also lists any chapters or subcategories which are not used, by title, followed by the note, "not used."

CONCLUSION

Every office should have an organized system for numbering drawings. This system should be used for every project. As shown in this chapter, some elements of the system can be omitted or expanded to fit individual project requirements, but the basic system remains unchanged.

DOCUMENT CONTENT BY PHASE OF WORK

8

INTRODUCTION

Architectural projects are traditionally produced in phases. These are:

Predesign
Concepts
Schematics
Design Development
Contract Documents
Construction Administration

Each phase has its own requirement for drawing content. For example, a concept drawing might be a "thumbnail" sketch of a dormer, a porch rail, or a room-relationship plan. It is seldom much more than an idea, a "concept." A contract document, on the other hand, must be explicit. It will contain exact dimensions and notes defining materials and finish.

This chapter deals with the kind of information needed and shown in each phase of work. It is organized from predesign through construction administration to demonstrate the building of information, one element on another, until the project is completed.

The Owner/Architect Agreement Will Affect Document Production

This book is about drawings and drawing production. In order to determine the amount of documentation necessary for a given project, it is necessary to understand the contractual relationship between the architect and the owner. Differing drawing requirements exist based on this relationship. Public work, for example, may have vastly different requirements than does private work.

Public versus Private Work

Public work often imposes the greatest restrictions on drawing production. The client may dictate drawing size, media of production, phase content, and bidding conditions, to mention a few. Private clients may demand far less of their architect. For them, it may be sufficient to produce minimal agency-acceptable drawings and the budget will match that minimum amount of effort. If architects attempt to produce full-service documents on a reduced budget, they would soon be out of business. Similarly, if architects produced minimal drawings for a public client, they could be subject to legal action for not completing their requirements.

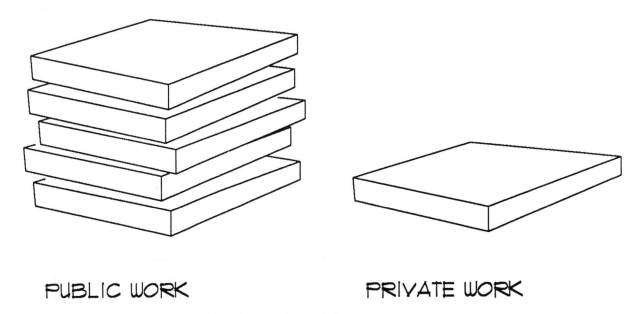

PUBLIC WORK   PRIVATE WORK

Figure 8.01 *The amount of documentation may vary with public and private work.*

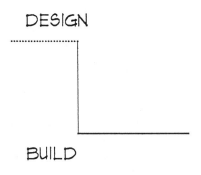

DESIGN

BUILD

CRITICAL PATH

Figure 8.02 Normal critical path.

Method of Construction Will Affect Document Production

Other considerations which affect the kinds of documents and their contents are fast tracking and design build.

Fast-Track Design

Fast tracking is a process which takes critical contract-document processes and relocates them to the design-development phase so the construction process can begin before the design is finished. This early package usually contains civil and structural drawings so that building foundations and framing can begin while the architect is finishing other contract documents.

Advantages and Consequences

The advantages are clear. When interest rates are high, a developer wants to speed up the construction process as much as possible so that tenants can begin renting in the shortest time. Another advantage of the fast-track process is the ability it affords to order long-lead-time items sooner. If, for instance, the contractor needs to order steel framing for a large project in a remote area like Alaska, the fast-track process can make the difference of many months of construction time. For the architect, it may radically affect the design-and-production process. The early ordering of steel requires the architect to have a stable design concept by the end of design development rather than being able to fine-tune design through contract documents. The client often feels the added design coordination effort is worth the benefits of early occupancy, and for the architect, satisfying a client could mean future business.

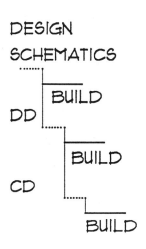

DESIGN
SCHEMATICS

BUILD

DD

BUILD

CD

BUILD

CRITICAL PATH

Figure 8.03 Fast-track critical path.

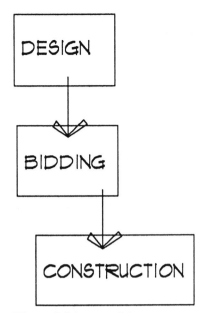

Figure 8.04 Normal design process.

Design-Build Process

The *design-build process* is a method of offering abbreviated architectural services without sacrificing overall building quality for the owner. The architect and engineering team produce only enough documentation to establish design and construction quality. The remainder of the work is provided by general contractors and their team of subcontractors. These contractors are brought aboard at project start-up, or in some cases even hired before the architect. The philosophy here is that each team member is responsible for her or his own area of expertise. The mechanical subcontractor, for example, will take preliminary performance-design drawings and specifications and prepare fabrication drawings and final engineering. This process eliminates the need for consulting engineers to do contract documents. The same process can be utilized for plumbing, electrical, and even structural systems. In prefabricated structures, the consulting engineer need only determine loading requirements. This is then passed to a pre-engineered building contractor for final engineering and building construction. In each case, the design professional must check the work of the contractor to protect the client's investment in their original design.

The Advantages of Design Build

The advantages of the design-build process to the owner are generally manifested in reduced design fees. This fact can be exploited by teams of architects and builders seeking new commissions.

Consequences to the architect arise when those reduced fees prohibit the development of ideas and designs. Often you must accept the way the contractor "estimated" the job rather than the way you would have designed it.

This chapter will deal mostly with a "normal" project from predesign through construction administration.

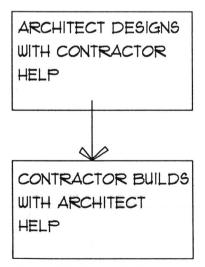

Figure 8.05 Design-build process.

PREDESIGN PHASE

DESIGN PROCESS

PRODUCTION STANDARDS

PROGRAM

EQUIPMENT COORD

FURNISHINGS SELECT

LAND USE

NEIGHBORHOOD ISSUES

BUILDING SURVEY

Figure 8.06 *Predesign is the time to ask critical questions.*

Evaluate the Project

Predesign is a time to draft contracts and road-map the process for producing drawings. It is the time to consider whether the project might be design-build or fast-track. It is the time to establish drafting and reprographics standards to be used in-house and by consultants and joint ventures. It is also the time to prepare the program, or if the owner has a program, to confirm its requirements. Predesign is the time to determine who will coordinate owner-furnished equipment (a very expensive and often-overlooked aspect of project planning). It is the time to identify potential problem areas like neighborhood committees against the project or land-use and code-related problems. It is the time to survey the existing conditions. And finally, it is the time to evaluate owners and their budget; can they pay their bills?

Get a Contract

Warning: The decisions made during the predesign phase will be binding throughout the life of the project. Time should be taken to study potential problems, evaluate alternatives, and make educated decisions. The people making the decisions must be familiar with design and production processes so that they don't "lose the farm" before the project even begins. Review previous projects with similar criteria, and check accounting records to determine profitability. When all looks good, then, and only then, sign the contract. What ever you do, don't start without a signed contract. Do no work without legal recourse for being paid.

With this behind, design work can begin.

DESIRED
DESIGN TRACK

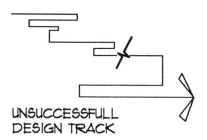

UNSUCCESSFULL
DESIGN TRACK

Figure 8.07 *The answers given at predesign can make or break the project.*

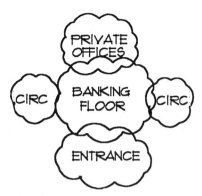

Figure 8.08 Block diagrams.

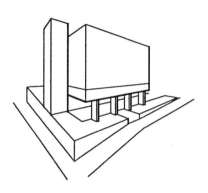

Figure 8.09 Models.

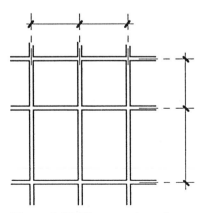

*Figure 8.10 Photographs and
sketches.*

CONCEPTS

Some projects, because of their size or complexity, require an early design phase called *concepts*. It is the purpose of this phase to begin the design and planning process in a manner which is preliminary even to schematics, and to establish the design language, building massing and systems, and to fix costs.

As the name implies, concepts are used to explore and record ideas. They are often not formal drawings, but rather vignettes used to describe elements of the project. A building concept can be explored in many forms. These include:

Block diagrams
Models
Photographs and sketches

Other formal drawings are seldom prepared at this phase of work with the exception of a site-utilization plan.

Block Diagrams

Block diagrams can be made which represent different departments and rooms, and then arranged to form various compositions. A good tool for recording the results is an instant-picture camera mounted on a tripod looking straight down. When a satisfactory arrangement is achieved, a picture is taken. Then other arrangements can be tried in the same manner.

If alternates are to be presented at the conclusion of concepts, change cameras and mount a single-lens reflex camera to the tripod. Take pictures of each scheme, and have a reprographics house blow them back on mylar. Using overlay techniques, combine the blowback "drawing" with project and office title block and informational notes. (See Chap. 6 for more information on overlay-drafting techniques.) The notes should include department and room names, gross square feet for each, and total gross square feet. A technique which adds interest and readability is to color code each department or room, especially when related rooms are not within the same major building block.

Models

Another tool used to study and present ideas is the *block model.* This adds the third dimension to flat-space planning drawings which helps define and illustrate massing. The process is begun by cutting and color coding wooden blocks to represent departments and rooms. Each block is made to be one story (about 12 feet) tall. For special areas like an auditorium, heights appropriate to the space are used. After the blocks are arranged, pictures are taken as with flat cut-outs. Additional shots are taken to show the third dimension.

If a more permanent method of displaying the concept massing is desired, a *massing model* can be made. This model is often made of high-density polyurethane foam blocks cut to the size and shape of the building. Test various schemes, up to three, and make a model of each. Use a single color for new construction with an alternate color for existing work. By using this technique, the model is more descriptive of massing than of departments or actual materials and finishes. Again, photographs can be taken to record the schemes and to use in promotional material.

Photography

Photography is often used for purposes other than just making a permanent record of models. In the realm of recording and illustrating architectural concepts, photographs can be used to show elements which might inspire design concepts. Field trips to buildings with significant architectural elements can yield many useful ideas. Other ideas can be obtained from photographs in books or magazines on architecture, travel, and nature. Photographs can also be used to record the existing conditions surrounding the building site, both for inspiration and to assist in designing an architecturally compatible building.

Sketches

The concept phase is used to study other items. For example, the program might require a "residential flavor." Thumbnail sketches which conjure up images of dormers, window seats, front-porch swings, and kitchen pantries can begin to describe a design philosophy for the project. If a "formal image" is requested, a sketch showing a structural grid might best set the design flavor. Each of these elements can be shown on a standard office drawing sheet and published with the massing plans to define the architectural concept.

Concept Site Study

At the concept phase a site drawing is often needed. Its purpose will be to study the ability of the site to carry the intended project. Information must be gathered about the site pertaining to zoning, fire, design review, historic preservation, geotechnical, covenants, and other factors which will influence and control the project. Items which can be mapped on the site plan such as building setbacks should be shown. This process will yield a graphic display of the buildable portion of the site. As a tool in the concept phase, the site drawing is used to help determine the feasibility of the project.

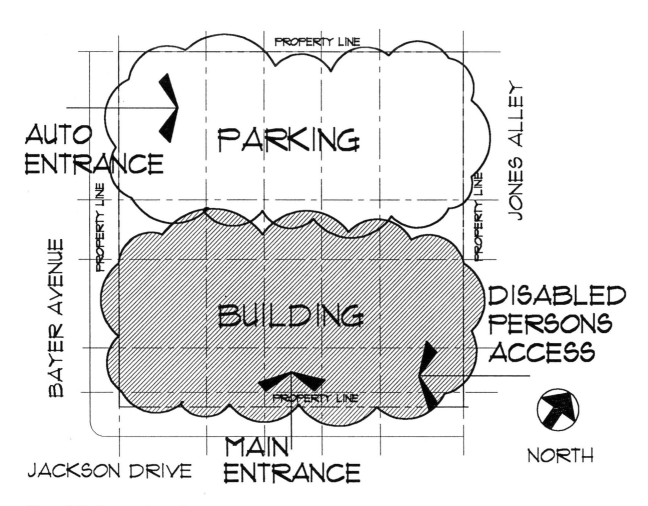

Figure 8.11 Concept site study.

Use Past Experience

Often, an architectural office is hired because of its design image. That image might be based on steel and glass or brick and wood. Along with those materials comes a history of idea development and problem solving. For example, if you use steel and glass for a multistory building, the problem-solving process behind the basic system includes items like heating and cooling needs. A mechanical engineer has joined the design team and helped determine glass area and even what kind of glass to use. This history of design and problem solving can be used to create new concepts in building design.

Keep Up with Technology

Another example for brick and wood designers is the new thin-wall masonry concept. By testing reinforced brick masonry walls under extreme conditions, new height-to-width ratios have become acceptable. Once a firm has added this type of information to its resource library, it can be used as an alternate concept for shopping malls or warehouses in lieu of tilt-up concrete construction.

Research Building Codes by All Team Members

The concept phase is a time to research and evaluate building codes, systems, and propose alternatives for the client's consideration. To do this professionally, it is important to have a team of consultants on-board and working. Don't wait until design development to bring the team together. Once the team is formed, don't let them sit back and wait for your input. They must all work together for best results. This will produce an integrated design.

Test the Concepts Relative to Cost

The concept massing and systems studies are sent to a construction-cost estimator to determine building costs. These estimates are used to steer the concepts in line with the owner's budget, or to help the owner determine a budget. When completed these items are packaged and dated under the title "Concept Documents." This package sets the design language, building massing, systems, and costs.

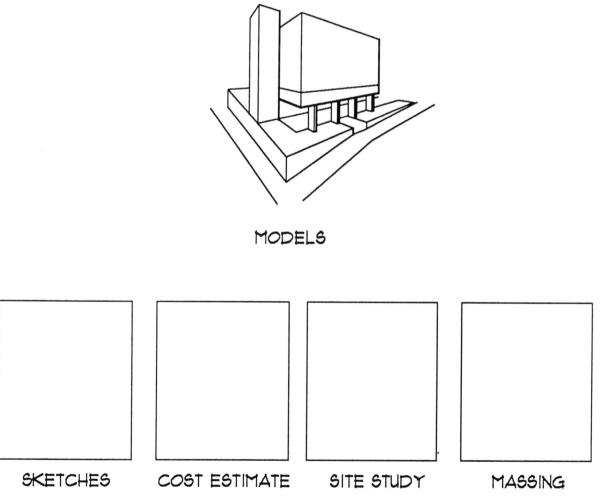

Figure 8.12 *The concept package.*

SCHEMATIC DESIGN

The *schematic-design* package is often the place where a project begins. Assumptions are made regarding building systems and cost based on past experience. If you have a good track record, this might be alright, but if you're new at a building type, it's best to start at the concept phase.

At schematics, the program for the schematic-design package is developed directly into room layouts. A series of drawings is begun showing plan, elevation, and section views of the building. The resulting drawings are packaged along with statements about building systems and cost control. The following is a more detailed study of what to include in a schematic package.

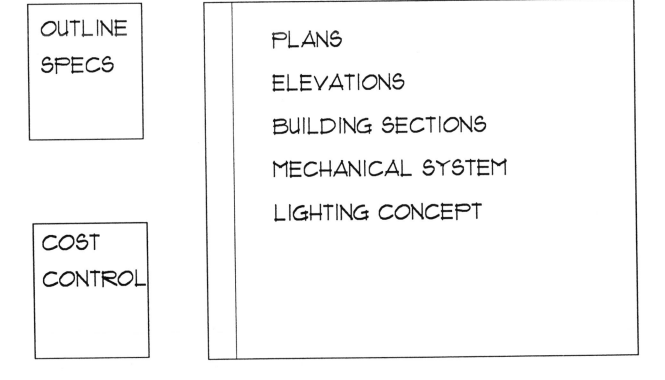

Figure 8.13 *Elements of a schematic package.*

Figure 8.14 Site plan.

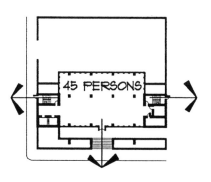

Figure 8.15 Floor plans.

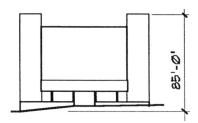

Figure 8.16 Building elevations.

Code Study

At schematics, a full-blown code study must be prepared. Begin with the site then determine:

    land use
    usable site area
    setback requirements
    bulk requirements
    parking
    street access
    any other pertinent information

Agency Interface

Contact the building officials and ask about their code requirements. Of special consideration are local environmental issues. Your state environmental policy act has provisions to protect the environment, and gives the public a variety of ways to either stop or delay a project. As the architect, a project plagued by continuous environmental reviews or design-review meetings can be financially dangerous. Production budgets can seldom absorb a lengthy code issue unless it is foreseen and budgeted-for early. Also ask the building official for any production standards. Some agencies require 24″-by-36″ sheet size and ⅛″ minimum lettering for microfilming clarity.

Drawings used to show the information gathered during the agency review are the following:

    site plan
    floor plans
    building elevations

The site plan will show zoning information for land use as well as building-code issues for yards and courts. The floor plans will show building-code information for exiting, occupancy-type and count, and for fire safety. Building elevations show massing and bulk. Once these items are established and approved by the building department, a record must be kept and used so the project does not swing off track.

THE DRAWINGS

Cover Sheet

The first drawing is the *cover sheet.* It contains the project identification, owner's name and address, and the list of drawings and credits. Additional information might include a vicinity plan so people can find the site. All of this should be drawn with the office standard cover sheet (see Chap. 1) showing the standard abbreviations and list of graphic symbols. If a prospective sketch is prepared, it may be placed along with project name and firm name on a precover sheet.

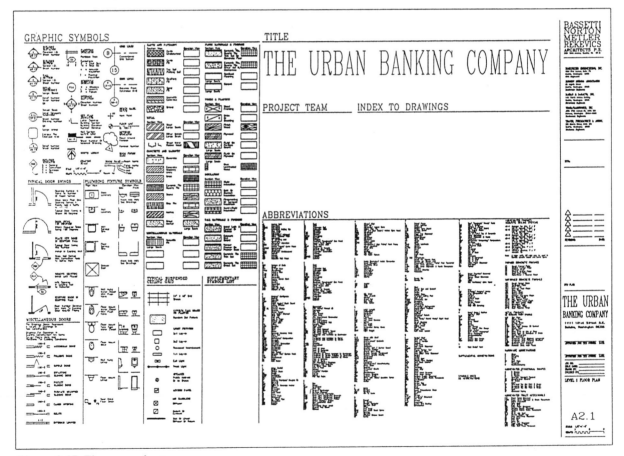

Figure 8.17 *The cover sheet.*

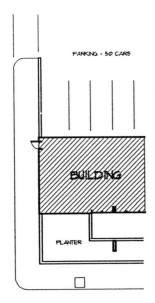

Figure 8.18 *Partial schematic site plan.*

Site Plan

The *site plan* will contain the basic information obtained at the concept phase including:

zoning
fire
design review
historic preservation
geotechnical and covenants

Some of this background data may be screened (see Chap. 6 on overlay-drafting techniques) to prevent it from obscuring the new information. Additional information obtained from the code study should also be added to the drawing, such as site access and setbacks. Show the building as a heavy line where the building comes out of the ground. Do not attempt to turn this drawing into a roof plan because you need to properly show what happens at grade. Show all pavement including street improvements, driveways, parking areas, walks, ramps, and site stairs. Locate existing utilities and proposed new utilities through the site. Show existing grades (screened for clarity) and the major concept for new grading. For schematics it may be enough to place arrows to denote slope. Show stormwater retention ponds. Obtain information from the civil engineer on size. Set the new building main-floor elevation. Show any existing structures, both to remain and to be demolished. Some agencies will require a planting plan to locate general landscaping improvements. Do not attempt a definitive landscape plan at this time. Note any other significant features such as a ravine or a bridge. Waterways draw particular interest from agency reviewers, even roadside drainage ditches.

This is one of the most important drawings to the building department when applying for site usage. Make sure all required information is shown. If necessary, divide the information into several drawings. Consider overlay drafting to accomplish this.

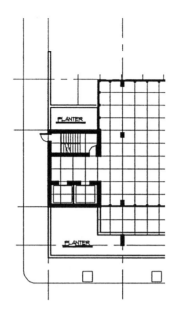

Figure 8.19 *Partial schematic floor plan.*

Floor Planning

From the program and adjacencies matrix, the *floor plan* is developed. At schematic phase, this should be at small scale, no larger than $\frac{1}{16}'' = 1'-0''$, and composed of single heavy lines. Wall thickness is not a major issue at this phase, but should be considered in tight spaces. Show the structural system of column grid and bearing walls. Work out to sufficient detail critical spaces like elevators, stairs, and toilet rooms. If adequate space is not provided at schematics, something else will need to shrink at design development. Save the larger-scale study sketches for technical planning later.

Show major architectural features of the building including:

roof overhangs
canopies
trellises
window openings
towers

Be sure the plan includes adequate space for handicapped persons, building maintenance, and for mechanical shafts. Show future expansion when it is an issue. Check corridor widths. Some building types require minimum corridor widths. Where there is no standard, use good sense and past experience to establish a width which is adequate and not wasteful. Make vestibules deep enough for the first door to close before the second door is opened.

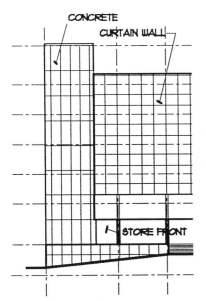

Figure 8.20 Partial schematic building elevation.

Building Elevations

These drawings are also produced at $\frac{1}{16}'' = 1'0''$ to match the floor plans. Their primary purpose is to convey the aesthetic design to the owner. The owner must be able to read these drawings and determine whether the design is acceptable. Good graphic skills are necessary to convey changes in plane and to show depth. The building elevations must also show building materials, fenestration, mechanical elements, and grade at the building line.

Building Sections

Draw at $\frac{1}{16}'' = 1'-0''$ to match the other schematic drawings. There should be two sections minimum, one longitudinal and one transverse. When existing structures are being added to, show vertical relationships. These sections should show the floor lines and roof line or top of parapet line. When the height is important because of zoning bulk requirements, show the reference lines defined in the ordinance. Show suspended ceiling cavities and dimension. Show the typical depth of structural members.

OWNER'S RESPONSIBILITY

The owner does not produce drawings, but does have responsibilities during the schematic phase. The initial space program should be expanded into a functional program to be used during design development. This is also the time to contract with interior designers and with vendors for major equipment. These people must be on-board during design development with the power of a contract sale so they can begin design work. If other consultants are not under the architect's contract, the owner must now bring them together. Another owner responsibility might be to make the land-use permit application. Some clients feel they can save money and time by doing this themselves. This practice is not recommended, however. Be certain your contract is not incumbered if permit delays occur.

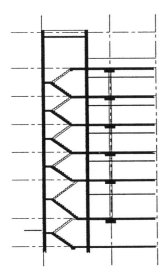

Figure 8.21 Partial schematic building section.

CONSULTANT'S DRAWINGS

Consultant involvement in every phase of project development is critical. Make sure they know they are needed by schematics and that their fee is to cover this time. They must be involved with program verification and at least oversee the events of other team members to help avoid problems. The best way to ensure this is at the time the contracts are negotiated with consultants. (See Chap. 6 for coordination of production methods and systems.)

Structural

At schematics, the structural engineer works very closely with the architect to establish the system of support for the building. Without this close working relationship, the design might never work. This is especially true when designing unusual buildings like inflated roof structures and tent structures. Some of the systems used are vendor-designed. One example is the space-frame structure. In this case, the vendor and structural engineer must work together to put together a whole building. Information generally needed at schematics includes:

foundations (estimated)
column spacing
bearing and shearwall layouts
floor framing concept
roof framing concept

Don't overlook the penthouse framing.
These items should be shown in the framing and foundation plans and building sections drawn at $\frac{1}{16}'' = 1'\text{-}0''$.

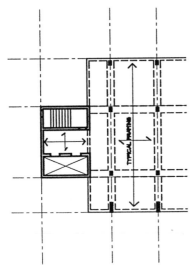

Figure 8.22 *Schematic structural drawing.*

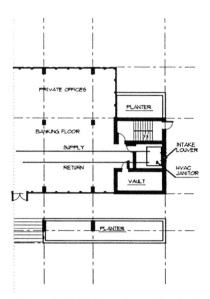

Figure 8.23 Schematic mechanical plan.

Mechanical

At schematics, the mechanical engineer should begin energy calculations to assist the architect with "what-if" problem analysis until a building shell is determined. Rough CFM requirements and air changes should be established. With this knowledge, the type of heating and air-conditioning equipment can be selected, and initial distribution systems determined. This should be illustrated on $1/16'' = 1'-0''$ schematic drawings so it can be seen that the system fits into the building. Size and weight of rooftop mechanical systems may have a large impact on wood frame structures. Early detection of these kinds of problems can save expensive and embarrassing redesign time. The mechanical engineer should also size water and sewer lines to the building and determine preliminary locations. These efforts will help produce a more meaningful cost estimate.

Electrical

At schematics, the electrical engineer should be able to determine basic building load requirements and size the kind of service entrance. From this, a location can be determined and a route determined for service across the site. Subpanel locations need to be established and closets provided. If the program did not state lighting requirements, the lighting consultant should now do so at each room type and on the site. A basic lighting allowance for energy-code purposes can be started to keep the building within allowance. Drawings should be at $1/16'' = 1'-0''$, and need to show service entrance, subpanels, and main conduit runs connecting panels to the main. Each room should show lighting allowance and any specialty features like location of auditorium lighting controls.

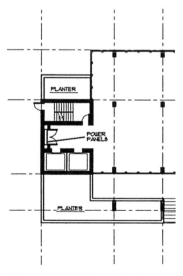

Figure 8.24 Schematic electrical plan.

Other Consultants

Some building types, because of their complexity, require special consultants to assist with planning. Some examples are sporting facilities, jails and prisons, and auditoriums. The schematic package from these consultants should show that the area they are consulting on will work as it is developed through later stages of work.

For example, a prison consultant must show a typical cell plan. An auditorium consultant must show a section view through the seating and stage, and an arena consultant must show the facility set up for hockey, the ice show, a rodeo, and other events. These drawings will be incorporated into the $1/16'' = 1'\text{-}0''$ scale schematic package.

CONCLUSION

The schematic phase is the time to put the program into three dimensions, and to study the resulting building as it relates to codes and the owner's budget. It is the time to begin to establish the systems of structure, heating, lighting, and power that service the building and make it stand.

Often the drawings produced are used to gain agency approvals for land use or other reviews. Be sure these requirements are incorporated. Another use of schematics is fund raising. Check the owner's requirements for architectural models and renderings.

Other documents that add value to the schematic package are:

Functional program
Area calculations
Code analysis
Project-development schedule
Cost estimate
Interior finish schedule by room type
Outline specification following Uniformat™
Owner-furnished equipment list

Have each consultant and the owner prepare their own documents for your coordination.

DESIGN DEVELOPMENT

Design development is the workhorse of the project phases. This is the time all major elements of the building are established. When this phase is completed, the project should be substantially designed and a final budget established. All that is left is to complete details and prepare final contract documents.

The key to a successful design-development phase is communications. Information is gathered and evaluated at a staggering rate, and the team needs to stay in tune with the project. Regular meetings need to be established (See "Quality Control," Chap. 10) to keep the design team tuned-up.

Each of the elements of design development are taken to the point that all pertinent questions are asked and answers received.

Code Study

The interface with local authorities is important throughout the project. Often new ideas emerge during design development which need code testing and agency interpretation. These items must be dealt with immediately to test their legality.

Start the process with schematic-approved designs and acceptable standards, and build on the relationship with the agency to gain favorable interpretations on new items. Do not change the design, building shape, or height in any way that will negate schematic-level approvals. This could result in the agency rescinding the previous approvals and requiring a complete and new application or review process. The results could delay a project by months.

Drawings at this phase should include the information issued at schematics for site development so that everyone can see there are no changes even though the drawing is beginning to look different. Add new information as it is discovered or required by agencies.

OUTLINE SPECIFICATIONS

At design development, an *outline specification* is prepared. It starts with a list in CSI (Construction Specification Institute) order of the divisions and subdivisions needed to define the project. Then, each subdivision is expanded to list the basic materials to be covered by that section and any pertinent preliminary items. The intent is to have a listing of every material needed on the job, but not to define it to the extent of a finished specification. The outline specification should have enough information for the cost estimators to fix the price of the building. Have each consultant prepare an outline for his or her areas of work.

```
EDMONDS - HAZELWOOD                                   SECTION 09660
ELEMENTARY SCHOOL                           .RESILIENT TILE FLOORING

                            SECTION 09660
                      RESILIENT TILE FLOORING


    PART 1 - GENERAL

    1.01  SECTION INCLUDES

        A.  Work includes but is not limited to following:
            1.  Vinyl composition tile laid over concrete substrate.
                a.  Colors and patterrns as indicated.
                b.  Provided by one manufacturer.

    1.03  SUBMITTALS

        A.  Submit samples including:  Standard covering samples, and
            terminations, for Architect's selection.

        B.  Submit Product and Maintenance Data in accordance with the
            requirements of Section 01730.  Provide three copies to
            Owner of product data and instructions on cleaning and
            finishing.


    PART 2 - PRODUCTS

    2.01  PRODUCTS

        A.  Vinly Composition Tile:  Tarkett,Inc., "Expressions
            Collection" and "Signals Collection", or approved.

        B.  Vinyl Reducer Strips:
            1.  Manufacturer:  Mercer Products Co., or approved.
            2.  Color as selected by Architect.
            3.  Location:  Provide transition/reducer strip at all non-
                flush floor transitions.  Coordinate with Architect.

        C.  Sealers, Fillers, Primers:  Water-resistant type, made or
            recommended by Manufacturer.
```

Figure 8.25 Example from an outline specification.

THE DRAWINGS

Cover Sheet

The design development drawings are often almost as numerous as contract documents. They therefore need a *cover sheet* with drawing index. Other information on the cover sheet should include:

> project identification
> owner's name
> owner's address
> credits for the design team
> a vicinity map so people can find the site

The architect may reproduce the rendering on the cover, but a disclaimer should be added noting potential inaccuracies in the rendering, and that it is not a binding element of the documents. This is important because many changes can occur in a building from the time of rendering and production of contract documents. If a cover sheet is produced with just the rendering and the project title, a second inside cover is needed to show the remaining items listed above and the standard office cover information (see Chap. 1) including standard abbreviations and list of graphic symbols.

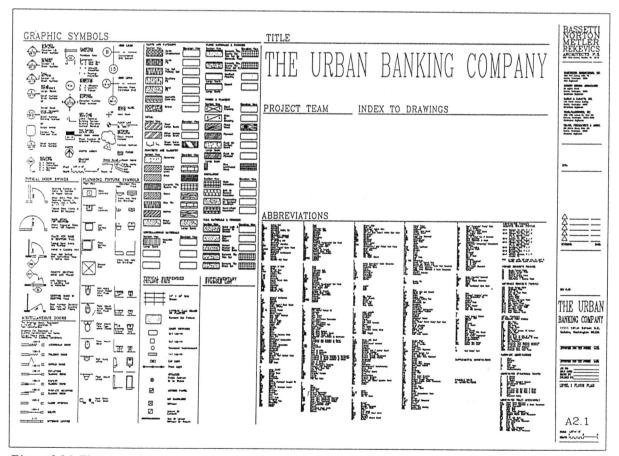

Figure 8.26 The cover sheet.

Site Plans

The *site plan,* for a site of any complexity, should be divided into civil, structural, mechanical, electrical, and architectural components.

Architectural Site Plan

The *architectural site plan* shows the footprint of the building (not a roof plan) and establishes dimensions to the property lines (see Chap. 2—"Dimensioning"); easements and setback lines and any other lines or boundaries restricting development; paved areas, including streets and roads, driveways, parking areas, patios, walks, and exterior stairs, indicating material as gravel, asphalt, or concrete; show existing grades as contour lines or spot elevations; set the new building by main-floor elevation; locate existing site features to remain, i.e., buildings, creeks, retaining walls, etc., and indicate such features to be removed; record underground features like buried service tunnels between buildings; show future expansion if this is a consideration; and miscellaneous site improvements like fences, benches, fountains, and sculptures.

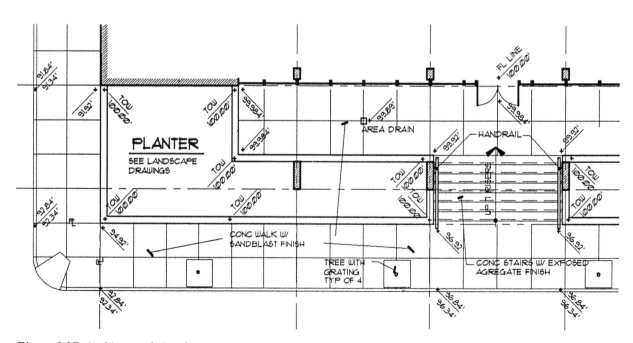

Figure 8.27 *Architectural site plan.*

CIVIL

Site Survey

The civil component may be shown on more than one drawing. The first is the *site survey.* This drawing is a record of conditions on the site and includes:

topography

trees and shrubs of significant size or species

existing improvements like walks, buildings, towers, and drainage ditches

survey of vertical surfaces of existing structures.

This drawing should be substantially finished at design development.

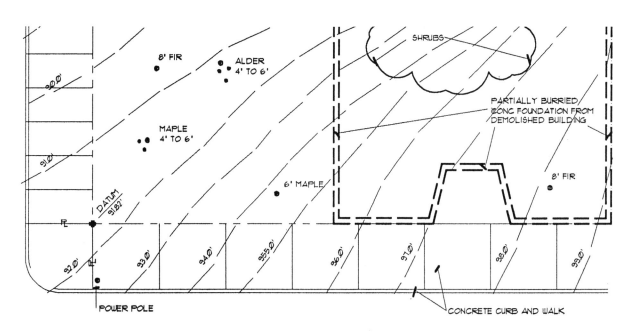

Figure 8.28 *Site Survey.*

Grading Plan

Another civil drawing will show *grading*. Contours are taken from the site survey and new grading patterns shown. This drawing will also show site utilities such as catch basins, yard drains, and oil separators. These will lead to a stormwater retention system, either on the surface or a buried tank. No detail work is necessary at this phase, but trench depths and back finning should be determined.

Paving Plan

A third civil drawing shows roads, drives, parking, and curbs. The drawings should note basic construction of each. No detail development is required at this time. This information is often shown jointly with the grading plan.

In rural areas, the civil engineer will design drain-field and septic systems. This is often shown on another unique drawing.

Structural

The *structural site plan* includes retaining walls and building footings. At design development, the footings might not be sized to their final dimension, and retaining walls will need further detail, but there should be enough information to establish the ultimate cost of the systems.

Landscape

A *landscape plan* will begin with the drawing started at concepts and schematics. For design development, they should include the names of plants and specific locations. Special grading requirements are shown for berms or other features. Show rock walls and stone paths which are not a part of either structural or architectural documents.

Mechanical

When no civil engineer is employed, the mechanical engineer will prepare the *site utilities drawing*. These include sanitary and storm sewer and water mains. By the design-development phase, these lines should be correctly sized and located on the site. This should include trench depths so that excavating and backfilling can be determined. Be sure the mechanical engineer is qualified to perform site work and realizes that responsibilities will include this work.

Electrical

The *electrical site plan* shows site lighting both pole-mounted and wall-mounted on the building. Respond to latest glare requirements. Show service to the building and transformers, both buried or on the surface. Show all wire runs and trenching requirements for proper cost take-off.

FLOOR PLANS

Composite Floor Plans

For large buildings, a *composite floor plan* is drawn at $\frac{1}{16}'' = 1'0''$. This composite has one main purpose: to show the entire building in one place. The building will be drawn at larger scale showing full detail development. The composite floor plan may be taken directly from the schematic drawing. Line work should be improved to include wall thickness and door swings. Each room should be named and numbered. The large-scale plans need to be referenced here by using break lines at drawing boundaries. This drawing should be substantially complete by conclusion of design development.

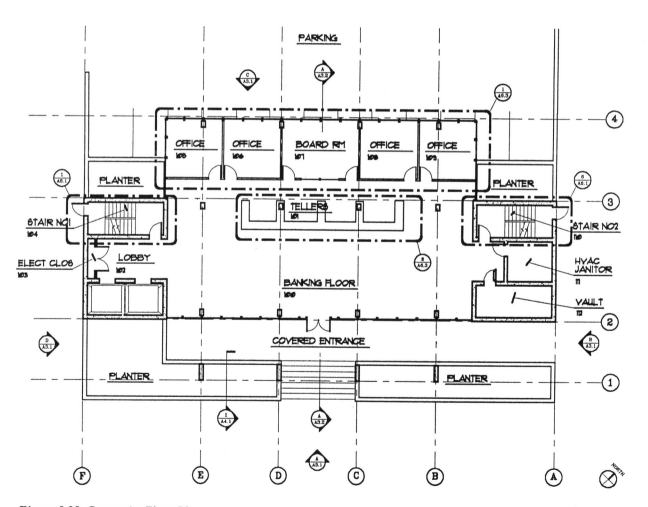

Figure 8.29 *Composite Floor Plan.*

Code Plan

The composite floor plan should be screened (see Chap. 6) and copied for use as the *code plan*. The purpose of this drawing is to show the following:

construction type

occupancy counts in assembly areas

occupancy counts at exits to justify the door sizes

building separations

area separations

occupancy separations

smoke separations

attic draft separations

rated walls for shafts and corridors

code notes

The intent of the code notes is to show how the building complies with the codes. This drawing should be substantially complete by the conclusion of design development.

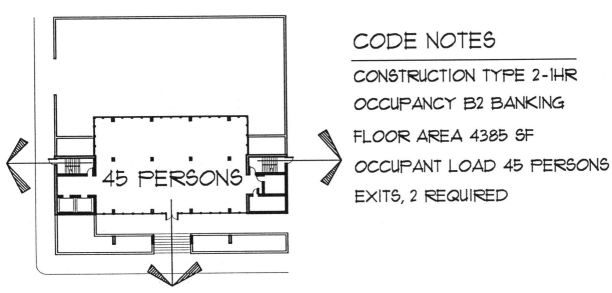

CODE NOTES

CONSTRUCTION TYPE 2-1HR

OCCUPANCY B2 BANKING

FLOOR AREA 4385 SF

OCCUPANT LOAD 45 PERSONS

EXITS, 2 REQUIRED

Figure 8.30 Code Plan.

Floor Plan

Floor plans should be drawn at ¼" = 1'-0". This will provide enough room for ⅛"-high lettering (often required by building departments), room titles, finishes, and miscellaneous notes. Show transitions between finishes, usually at doorways, but anywhere they occur. The "typical" partitions for nonrated, 1-hour, 2-hour, and shaft walls should be determined. They can be located from the Code Plan, so no referencing is required at this phase. Identify nontypical partitions and furring. Show all material poché, i.e., concrete, brick, and other masonry units. Dimensioning should be done only for building exterior elements, structural system, and other critical places (see Chap. 2—"Dimensioning"). At design development, the kind and size of equipment is important (see Equipment Schedule in Chap. 3). Every item in group-I equipment and some group-II items with space requirements must now be shown. Items furnished and installed by owner are drawn with a dotted line. Items furnished by either the owner or contractor and installed by the contractor are shown by solid line. Each item is numbered on the plans.

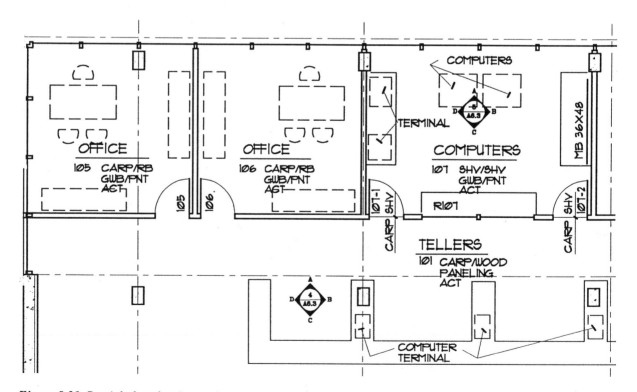

Figure 8.31 *Partial plan showing equipment.*

Each door, relite, and other opening which receives hardware is numbered on the plans (see "Door and Openings Schedule," Chap. 3 and "Graphic Symbols," Chap. 1). New doors are shown with 90-degree swings unless a full 180 degrees is required. Existing doors are shown with a 30-degree swing. Door numbers for existing doors requiring no work may be omitted. Cabinetwork is shown by drawing countertops and splashes with a solid line. Upper cabinets are drawn with a dashed line. All cabinet depths will be assumed to be standard 24" for bases and 13" for uppers unless dimensioned here on the plans. Indicate wall elevations with casework configurations. (See "Graphic Symbols," Chap. 1.) Do not note casework configurations or dimensions on the floor plans (see interior elevations). Show chalkboards, marker boards, tack boards, motion-picture screens, and similar items. Show room services such as hot and cold water outlets, even when there is no sink, waste receptacles, electrical outlets, and communications jacks. Show only enough detail to clearly define the project at this phase. If, for example, a large equipment item will require special floor reinforcing, note the requirement on the plan but do not produce the detail.

When the building size does not require a composite to show the whole plan, then consider the plan at $\frac{1}{8}" = 1'-0"$. When this is done, the information shown on $\frac{1}{8}"$ and $\frac{1}{4}"$ drawings is split. The $\frac{1}{8}"$ drawing should show all material poché for concrete and masonry, room names and finishes, dimensions for exterior and walls and interior partitions, and door numbers. The $\frac{1}{4}"$ plans should then contain casework and equipment data. Do not draw the entire building at both scales. If there is a lot of casework or equipment, consider drawing only $\frac{1}{4}"$ plans. If there is little of this, perhaps $\frac{1}{8}"$ plans will suffice. When in doubt, choose the larger-scale plans. It may take more sheets, but they will be much easier to produce. If $\frac{1}{4}"$ plans are drawn, and the entire building cannot be shown on one plan, revert to $\frac{1}{16}"$ or $\frac{1}{8}"$ for a composite, and follow the descriptions above.

Building Elevations

The design-development *building elevations* should be drawn at the same scale as the floor plans. Choose ¹⁄₁₆″ when composite plans are drawn at ¹⁄₁₆″, and choose ⅛″ when basic plans (or composite plans) are drawn at ⅛″. (See also Chap. 2, "Drawing Scale.") The building elevations are the place to show all building exterior materials and finishes. There is no equivalent to the Room Finish Schedule for exterior spaces, so all expressions of material and finish must be shown on the elevations. Also shown are windows, curtain walls, doors, louvers, and architectural elements like canopies, exterior stairs and rails, ramps, overhangs, and roof edges. At windows and storefronts, reference the Window Schedule for more detail information. Reference only details that establish design. Show floor lines and vertical openings (see Chap. 2, "Dimensioning"). Show mechanical elements such as stacks and exposed cooling towers. Show grade lines, both new and existing. At conclusion of design development, the building elevations should be substantially complete except for detail referencing.

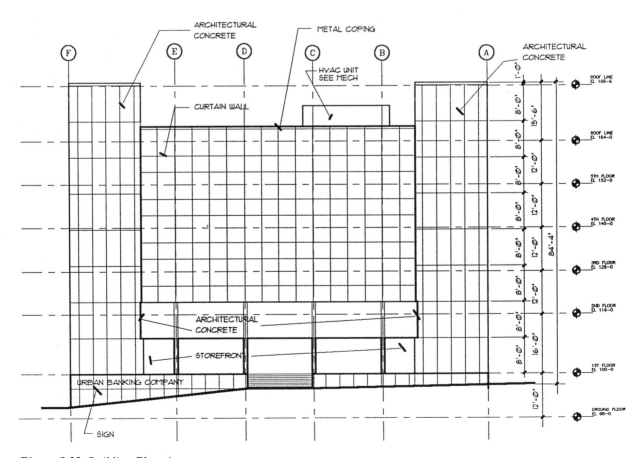

Figure 8.32 Building Elevation.

Building Sections

Draw at the same scale as building elevations (see "Selecting Drawing Scale," Chap. 2). This drawing shows floor lines and column grids and vertical dimensions along the outside edges of the drawing (see "Dimensioning," Chap. 2). Show structural elements including beams, joists, girders, and trusses. Note general fireproofing requirements for floor-ceiling assemblies and for roof-ceiling assemblies. Do not show wall elevations; however, do show exterior elevations when they occur as part of a building section. Show rooftop elements like penthouses and cooling towers. Draw at least two sections, one transverse and one longitudinal. Avoid stair and elevator shafts as these are shown at larger scale. When this drawing shows a portion of the building in elevation, follow the description written for building elevations. This drawing should be substantially finished at the end of design development.

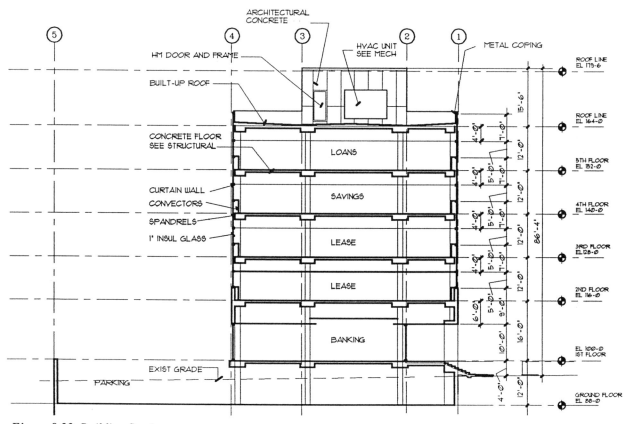

Figure 8.33 *Building Section.*

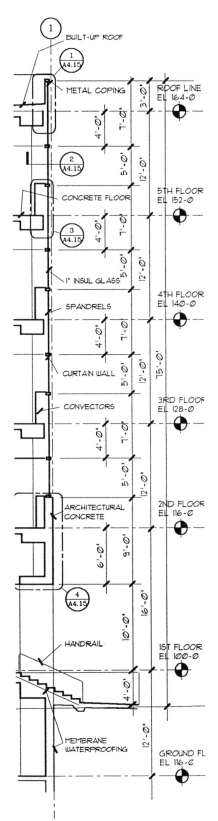

Wall Sections

Wall sections are drawn at ⅜″ = 1′-0″ to ¾″ = 1′-0″. (See "Selecting Drawing Scale," Chap. 2.) Wherever possible, coordinate with structural drawings. These drawings should start with the "typical condition" in the building. For example, if the building is made of brick veneer over wood framing, draw that condition. Do not include windows, doors, louvers, or other openings. Just draw the typical wall condition. Follow this by any other sections needed to define wall conditions at windows, doors, and other conditions. These conditions can be for both exterior and interior walls as the case demands. The building could have a number of typical sections. Draw each one after exploring variations on the first one. When all typical conditions and their variations are drawn, draw the special conditions. This includes the condition against existing structures. Show column lines and building lines for vertical reference. If the building facade is very three-dimensional, show the dimensions relative to the building line. Draw sections through windows and storefronts and curtain walls. Note materials and finishes, and dimension critical elements. Much more detail will still be needed during contract documents.

Figure 8.34 Wall section.

1'-6"

BRASS REVEAL

1'-6"

OAK VENEER
PLYWOOD

3'-6"

OAK TOEBOARD

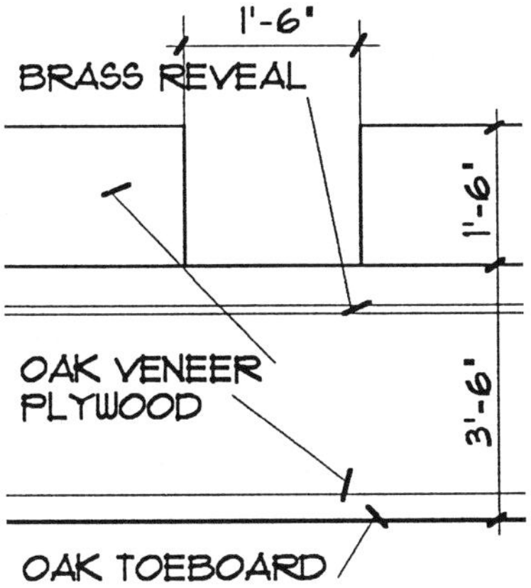

Figure 8.35 Interior Elevation
showing casework.

Interior Elevations

Interior elevations are drawn at $\frac{1}{4}'' = 1'-0''$. (See "Selecting Drawing Scale," Chap. 2.) The purposes for these drawings are many. This is the place to show wall finishes, especially when more than one condition exists on a wall. For example, a wall could have different materials and finishes on a pilaster. The Room Finish Schedule (see Chap. 3) cannot show this, so the elevation must.

Show architectural features such as pilasters and soffits, and trim items including baseboard, chair rails, corner guards, and cornices. Show casework, cabinetwork, millwork, doors, relites, and any other feature. In toilet rooms, elevate fixture walls and walls receiving accessories like paper-towel dispensers and waste receptacles. At cabinetwork, show inside shelving (dotted), light valances, and all plumbing fixtures, both deck-mounted and wall-mounted. Do not show cabinet door swings but indicate swing by placing a handle in the appropriate location. Also show tack boards, chalkboards, marker boards. When equipment is mounted on the wall, show the item as on floor plans, dashed lines for owner-installed items and solid lines for contractor-installed items. Dimension spaces, both vertical and horizontal, for items of known fixed size, like a refrigerator, and say, "HOLD." Size cabinets by running a string of dimensions along the bottom of the drawing and along one side noting the sizes in inches. (All depths will be standard unless noted on floor plans.) At design development, no further detail is necessary except as may be used to explain a design feature, i.e., crown molding.

Reflected Ceiling Plans

Drawn at the same scale as the base floor plan (⅛″ or ¼″ = 1′0″.), this plan is often under fire as unnecessary and a waste of good production dollars, especially in developer work, but I endorse it as always important and especially in developer work. The main purpose of this drawing is to act as a referee among the trades by locating in advance the positions for light fixtures, ceiling diffusers and, exhaust grilles. As a design tool, it is used to show architectural elements like soffits, crown molds, vaults, skylights, and more. This plan should show everything which is visible on the ceiling. In most work, the ceiling surface is flat and level, or it follows the slope of the structure to which it is attached. For level ceilings, note the height in a box as "X" feet and "Y" inches above reference floor line. Note other planes in similar fashion with a box and their height above the reference floor line. Sloping ceilings need elevation marks at high and low spots much like a roof plan. These areas are often best-described in building section view, especially for complex ceilings like in an auditorium. Like all drawings, note material and finish. Coordinate with the Room Finish Schedule. For complex ceilings, such as in an emergency room at a hospital, the ceiling plan should be drawn at ½″ = 1′0″ for greater clarity. Show only enough detail to establish the scope of work and fix the budget.

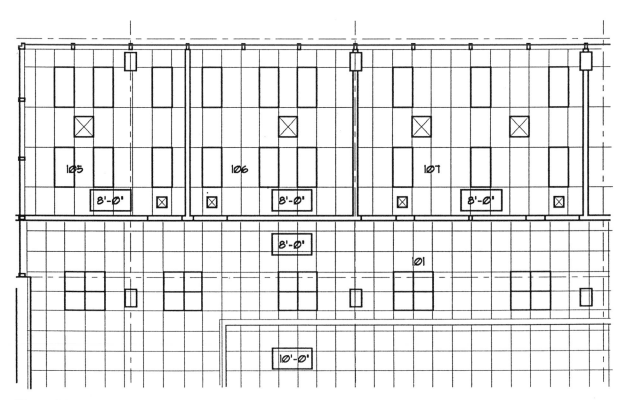

Figure 8.36 Reflected Ceiling Plan.

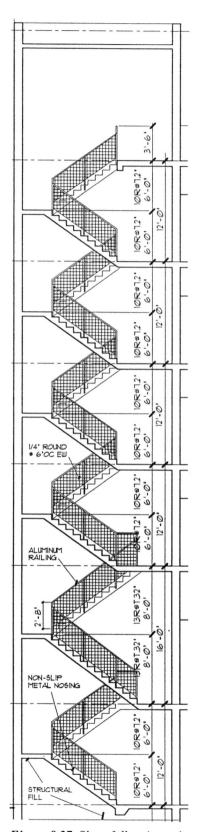

Figure 8.37 *Show full stair sections.*

Vertical Circulation

Take the study sketches prepared at schematics and develop full plans and sections for all stairs, ramps, elevators, and escalators. Elevators and escalators may be long-lead-time items, and are greatly affected by building, handicap, and life-safety codes. Check the codes and determine their requirements. Make a list and check it off as you develop the systems. Develop vertical circulation in plan and section at the same time on the same drawing sheet. For stairs, show the entire shaft, not just stair elements. This includes all floor-framing members that run through the shaft (code permitting) to verify head clearance. Show rise and run on the section view of the stair run nearest you, and allow the other (switch back) run of stairs in elevation. (See "Dimensioning," Chap. 2.) Note special nosing requirements. Show handrails and rail extensions. At elevators, pay close attention to codes. Check shaft size, pit depth, and elevator overrun at penthouse. For escalators, be sure there is enough underfloor space for hidden machinery. Prepare only enough detail to determine the systems work and meet code. When unusual design issues are involved, draw enough detail to show shape, material, and major supports.

Schedules

At design development, begin preparing all schedules. On most forms, the data needed will include "what" and "where" type of information. For example, on the Door and Opening Schedule, give door size, configuration, frame type, and general hardware requirements. On the Room Finish Schedule, list all materials and finishes. On Window, Louver, and Panel Schedules, show all shapes, materials, and finishes with overall dimensions. On the Equipment Schedule, fill in everything except specific mounting details. If a Casework Schedule is used, identify the casework and accessory items and produce the typical cross sections. In general, fill in schedules with as much information as you can, excluding the construction details.

Details

The design-development phase is not the time to completely detail the building; however, some detail is needed to define shapes, establish design intent, and give the cost estimator something to work with. This is especially true as the ornamentation of a building increases. When details are needed, do not attempt to solve all the construction problems. Go far enough to know that the condition works, or that a "typical" condition works, and leave the rest for contract documents. This is very important. Study the conditions far enough to *know* they will not substantially change during contract document phase. Produce details using the detail-book file system (see Chap. 7) so they can be found for later development.

INTERIOR DESIGN

The building type will have a strong influence on when you bring in the interior designer. A large hotel or home-office building may require strong interiors work during design development. An industrial project might need minimal interiors work, and only during contract documents. Based on program needs, the interior designer can work with the owner and architect to produce an integrated building. There are often arguments over who has the lead in interior architecture. It is best to discuss these items at length before anyone begins work. Define the exact degree of involvement by the interiors group in items like space planning. Do not allow the interiors group to deal directly and alone with mechanical, electrical, or structural engineers, even if they are working for the owner and not for the architect. As the architect, you must coordinate these items, usually in session with interiors and other consultants. The interiors documents are often the same plans and wall elevations used by the architect, copied, renumbered, and modified to show finish requirements. At design development they need to show specific locations of nonstandard items for cost-estimating purposes.

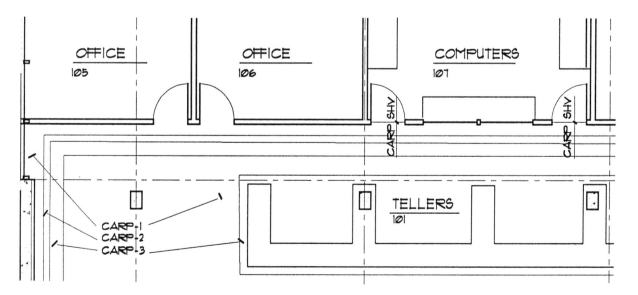

Figure 8.38 Interior work is often shown over architectural backgrounds.

OWNER'S RESPONSIBILITY

As in schematics, the owner also has responsibilities to the project. The three biggest jobs for the owner are:

meeting financial responsibilities
completing equipment lists
meeting with the architect
contracting for outside services

These meetings with the architect are sessions aimed at establishing the finite requirements of the building. Since it is generally the owners who pay the bills, their economic health is the architect's business. The project cannot live without it. The architect has the right to know that he or she will get paid and toward what building budget they are working. The owner must complete equipment forms and sign contracts with all vendors which were not signed during Schematics. The information they have to contribute at this time will put the job behind schedule if allowed to trickle in later. Insist on this information now. Finally, the owners and their staff need to dedicate many hours of time to meet with the architect to discuss, in conferencing sessions, the fine points of their needs. These meetings start with the space program and functional program and begin to put the written word into three dimensions. Close owner input is necessary to design a functional building. Once the design development has been completed, have the owner sign off and accept the work.

Figure 8.39 Get the owner's signature of approval.

CONSULTANT DRAWINGS

The consultant's design-development documents must keep pace with the architect's. There is often the argument that consulting engineers cannot design until the architect stops changing the plans. This problem will not arise if everyone works together, owner, architect, and engineer, to reach a finished design development together. This joint conclusion of the design is important to project schedule, and must be given due consideration. The architect who cannot stop designing will force the consultants to do their designing during contract documents, and the owner will pay through change orders.

Mechanical

The energy analysis should be finished and coordinated with the architect. All mechanical plans should have ceiling diffusers and return/exhaust grilles located. Duct runs should be firm and drawn as a one-line diagram at the same scale as the basic architectural floor plan. (See "Production Systems," Chap. 6.) All mechanical equipment should be located. Major items like HVAC (*Heating, Ventilating, and Air-Conditioning*) units should be sized and specified for coordination by the other members of the design team. A rooftop unit can have a strong visual impact if not properly integrated into the design. Smaller items like pumps and exhaust fans might only be located so the space considerations are addressed. Fixture cuts of all units, sinks, faucets, fans, grilles, etc. need to be assembled for the architect's coordination. Plumbing drawings need to show mains and risers so furring can be provided and coordinated. When fire-sprinkler systems are installed, the plumbing drawings need to show the size of the main, location of the shut-off valve, and pressure available at the valve. The rest is usually engineered and built by a sprinkler contractor. If possible, have that person on board and working.

Electrical

From the conferencing sessions with the owner and information shown on the architectural drawings, the electrical engineer can determine the building's power needs. All main and subpanels should be located with feeders connecting them. The transformer should be sized and located. A lighting plan should be drawn over the architect's background (see Chap. 6, "Production Systems") and fixture cuts provided for all types, including exit signs and emergency path lighting. Rooms with special lighting demands should have a substantial amount of engineering completed at the end of design development. This includes:

> power-distribution centers
> auditorium lighting
> power demands for heavy machinery

Other Consultants

As design development progresses, the specialty consultant must complete her or his work. The assistance the consultant brings to the design should be concluded with the end of design development, or the design isn't really finished. Work closely with the specialty consultant to get all the help you can.

CONCLUSION

The design-development phase of any project is the time to define the work, not in detail but to a point that the design is substantially complete. The documents should all be complete enough that a whole new crew of architects and consultants could pick it up a year later and finish contract documents without meeting with the owner to ask questions about missing information. The reason is that this very thing sometimes happens, and there isn't always enough fee to go back and have the owner fill in the empty blanks. Take the time to prepare outstanding design-development documents, and the remainder of the project will run much smoother. Do not proceed until the owner approves the package.

Other documents that help define the building or back up the drawings and specifications are:

Area tabulations

Code notes

Copies of correspondence with the building official

Project-development schedule

Cost estimate

Equipment list

Book of equipment data sheets

Soils analysis

Energy statement

Handicap analysis

Acoustic analysis

Include any special report that will help define or back up the building design.

CODE ITEMS

SITING

CONSTRUCTION TYPE

OCCUPANCY

EXITING

FIRE PROTECTION

ASSEMBLIES

HAZARDS

LIFE SAFETY

Figure 8.40 *Study the building codes.*

CONTRACT DOCUMENTS

With most of the design work already finished, *contract documents* is the time to solve the technical problems and to finish the drawings, schedules, specifications, and other documents needed to obtain a firm cost and to build the building. The success of the project hinges on working toward these goals and not backtracking into the design-development mode.

Following is an outline of documents needed for most building types and a description of what they include.

CODE STUDY

Major issues of zoning and building should already be determined. At contract documents, code issues should center on the technical issues like corridor construction and brick veneer detailing. As each condition is discovered, discuss it with the building official and get an interpretation on any questionable issues. Do not assume, for example, that the airspace behind veneer masonry can be less than 1″ just because you didn't think of it during design development. Once the building frame is constructed, the building official may require that the face of brick be moved outward.

One source for assistance in solving code-related problems for people in UBC country is the International Conference of Building Inspectors (ICBO). As members, architects may call for interpretations. (*Caution:* the building official may have the right to rule differently from ICBO. Check before making assumptions.) ICBO also publishes the UBC *Standards* to guide the architect in technical problem solving.

Agencies often require that material assemblies be tested to ensure their fire resistance. There are a number of testing laboratories that publish booklets on tested assemblies for floors, roofs, walls, and partitions. Another publication, *The Gypsum Association Handbook,* lists fire and acoustic ratings for gypsum board on wood or metal framing for walls, partitions, and furring.

AGENCIES

BUILDING OFFICIAL

FIRE MARSHAL

THE CLIENT

BOARD OF HEALTH

OSHA

EPA

FAA

and many more...

Figure 8.41 *Sources for code information.*

A group of codes often referenced by local authorities is published by the National Fire Protection Agency (NFPA). These thirteen volumes of 100-plus codes deal with every fire condition imaginable, from 20-minute corridor doors (NFPA 80) to the Life-Safety Code (NFPA 101).

Each specific building has its unique set of code requirements, and proper technical problem solving can not begin until each and every applicable code has been identified and studied. It is the architects' responsibility to know what codes affect every project they design, and how the codes apply to each situation. It is not enough to say that contractors are also bound by the codes, so they can catch the problems. If you don't know what codes apply, ask. Some agencies to ask include:

The building department
The engineering department
The fire department
The health department

If the project is a public job, check state and local rules. Ask the owners, they often know. Keep the latest edition of the most common codes in the technical library, and read them often enough to stay familiar with their general requirements.

CONTRACT SPECIFICATIONS

It is good for every architect to write a specification now and then, and the process of writing the design-development specification should be conducted by the project lead technical architect. But, at contract documents, it is better to place this task in the hands of a specialist. If your practice is large enough, you might assign an individual to write specifications, but if it is not, there are outside consultants who can prepare a very good document, and likely in far less time and for far less money than can be done in-house. Seek out a Specifications Writer who can offer a variety of formats from outline to full-blown CSI. Find one with a varied price range to match the extent of the job. Work with the writer to coordinate the information that is received and published to make sure that all documents are compatible.

```
EDMONDS - HAZELWOOD                              SECTION 09660
ELEMENTARY SCHOOL                       RESILIENT TILE FLOORING

                          SECTION 09660
                      RESILIENT TILE FLOORING


PART 1 - GENERAL

1.01  SECTION INCLUDES

   A.  Work includes but is not limited to following:
       1.  Vinyl composition tile laid over concrete substrate.
           a.  Colors and patterrns as indicated.
           b.  Provided by one manufacturer.

1.02  RELATED SECTIONS

   A.  Coordinate related work specified in other parts of the
       Project Manual, including but not limited to following:

       Section 03300   -   Cast-in-Place Concrete
       Section 09651   -   Cementitious Underlayment:  For
                           leveling floor surfaces
       Section 09678   -   Resilient Base and Accessories
       Section 12300   -   Manufactured Casework

1.03  REFERENCE STANDARDS

   A.  Comply with the requirements of Section 01091 and as listed
       herein.  See Section 01091 for listed association, council,
       institute, society, and the like organization for its full
       name and address:

       American Society for Testing and Materials (ASTM):
       ASTM E84-87 Test Method for Surface Burning Characteristics
                   of Building Materials.
       ASTM E648-86    Test Method for Critical Radiant Flux of
                       Floor Covering Systems Using a Radiant Heat
                       Energy Source.
       ASTM E662-83    Test Method for Specific Optical Density of
                       Smoke Generated by Solid Materials.
```

Figure 8.42 Specification writing is more technical than many architects are capable of producing.

THE DRAWINGS

Cover Sheet

Reuse the cover sheet from design development. Change the phase name and date, and revise the drawing list. Verify all other information as still accurate or revise. This sheet should not need much more work.

Site Plan

Reuse the design-development drawings. Update the elements shown if any changes have occurred. Add dimensions for all pavement areas and reference all site details. (See Chap. 7—"Document Identification and Order" for the place to show these details.) The same applies for civil, structural, landscape, mechanical, and electrical site drawings. The basic scope of work should have been established during design development, so now is the time to draw details and fix dimensions. Be sure there is a project sign and location selected stating the names of the project, owner, architect, contractor, and all sublevel professionals. Show contractor staging area and limits of construction and construction access when appropriate. Caution about limits of construction: Be absolutely sure there is no work outside the limits line. Often there are utility lines which extend beyond the "General Limits of Construction," and the contractor may claim an extra cost for doing that work.

On the grading plan, show area for stockpiling topsoil for redistribution at finish grading. On other civil drawings, look for details on manholes, catch basins, driveway construction, curb edges, and septic system.

On landscape drawings, show all planting, especially that which is required by local agencies. Detail tree and shrub planting and staking. Detail the irrigation system. Detail features not otherwise covered on the architectural or structural site plans, i.e., site benches, berms, planting edges, fountains, and the like.

On the electrical site plan, show site lighting details at grade and on buildings. Detail standard trench and backfill.

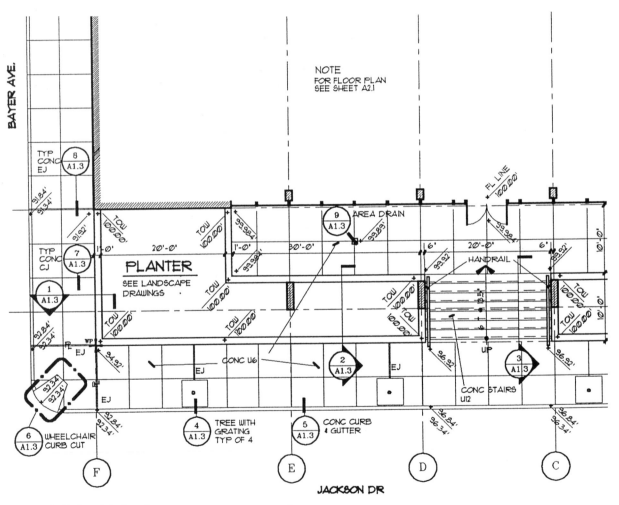

Figure 8.43 Partial site plan.

Floor Plans

The composite floor plan should be substantially complete at the end of design development. Check both for changes and update. Reference all large scale plans on the composite floor plan. Show north arrow.

On the basic plan, at $\frac{1}{4}'' = 1'\text{-}0''$, complete dimensioning (see Chap. 2, "Dimensioning"). Do not overdimension. Allow column centerlines, window centerlines, and other features to set the locations for walls and partitions where possible, and don't close dimension strings for interior partitions. Indicate partition and furring types with their partition symbol (see Chap. 3, "Partition Schedule"). Update all equipment. It is difficult to get all equipment data during design development, so check now to see that all contractor-installed equipment is shown and mounting details drawn. Coordinate with the structural engineer for heavy objects mounted on walls or suspended from ceilings. Reference the details on the Equipment Schedule (see Chap. 3, "Equipment Schedule"). Doors, relites, and other openings should be already shown and numbered. Check that any additions are shown on the schedule (see Chap. 3, "Door and Opening Schedule"). Name and number each room and note the key for interior finishes if the Schedule of Finishes is used (see Chap. 3, "Room Finish Schedule"). Show cabinets and countertops with splashes as solid line and wall cabinets as a dashed line. Show stairs with nosing solid and a second line 1" back for the toe space. This helps people see which way is up. Show plumbing fixtures and check with plumbing drawings. Do not show a tank-type toilet if you are actually getting a wall-hung tankless unit. Show fire extinguishers, fire-hose cabinets, and stand pipes. Conclude by writing general notes to help the contractor understand the set (see Chap. 9, "Finding Your Way Through the Drawings"). Cross reference all drawings accordingly.

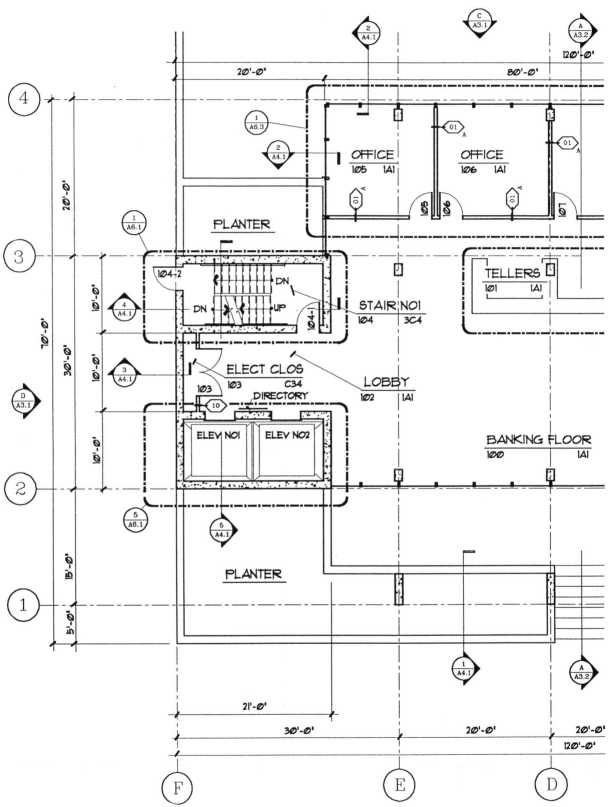

Figure 8.44 *Partial Floor Plan.*

Roof Plan

The *roof plan* is usually the last in the series of plan sheets, however, for multistory buildings with roofs and rooftop gardens adjacent to occupied space, the roof plan will be part of the floor plan. Either way, the roof plan shows roof-cover material, slopes, and drainage devices. Slopes are shown by arrows and by spot elevations at high and low points and along perimeters at copings, gravel stops, and penthouses or other walls (see Chap. 2, "Dimensioning"). Show equipment curbs, skylights, and hatches and detail them. Show rooftop mechanical units as dashed lines so the drawing more clearly shows the roofing conditions.

Show roofing expansion and building expansion and earthquake joints. Show roofing vents when required. Locate walking pavers. When window-washing equipment is present, show tracks, davits, etc., and details. Show ladders on the building exterior and detail. Show screen walls around mechanical equipment. Check airspace and clearances. Detail walls, especially at roof penetrations. At open stairs to roofs, check code requirements for guardrails and detail (see Chap. 9, "Finding Your Way Through the Drawings," and cross reference all drawings accordingly).

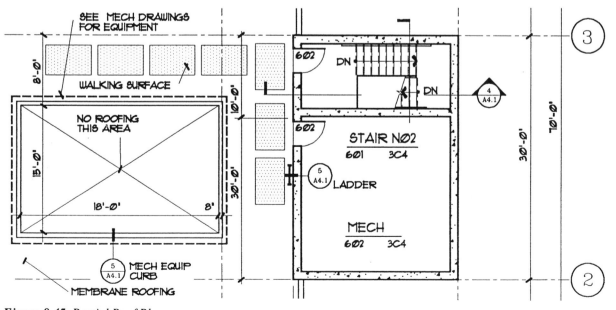

Figure 8.45 *Partial Roof Plan.*

Building Elevations

The design-development building elevations should be substantially complete. Check for all material designations and window, storefront, and curtain walls references (see Chap. 3, "Window Schedules"). Where large-scale elevations are drawn, show the reference (see Chap. 9, "Finding Your Way Through the Drawings"). Elevate all exterior surfaces, even those inside exterior screen walls on roofs. This is the only place to describe exterior materials and finishes. Show downspouts, expansion, and earthquake joints and detail. For cast-in-place concrete buildings, show form tie locations if the appearance is important. Show wall-panel joints in concrete, stucco, and other panelized systems. If a Panel Schedule is prepared, cross reference here. Show exterior ladders at penthouses. Show mechanical items, especially large fans and cooling towers because of their aesthetic impact as well as any height impact relative to zoning. Show both existing and finished grade lines.

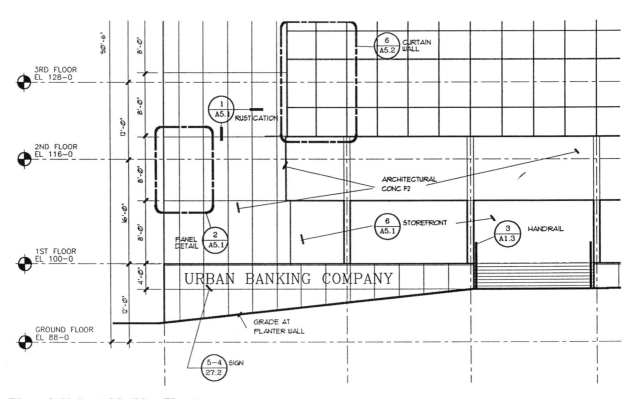

Figure 8.46 *Partial Building Elevation.*

Building Sections

The design-development building sections should be substantially completed. It may be necessary to draw additional sections at contract documents to clarify conditions. Prepare them in the same manner as noted previously. When building elevations are combined with building sections, show items noted for building elevations. When an existing building is being expanded, consider several partial sections at the joint. Draw enough to make the condition clear, especially when there are changes in floor or roof elevation.

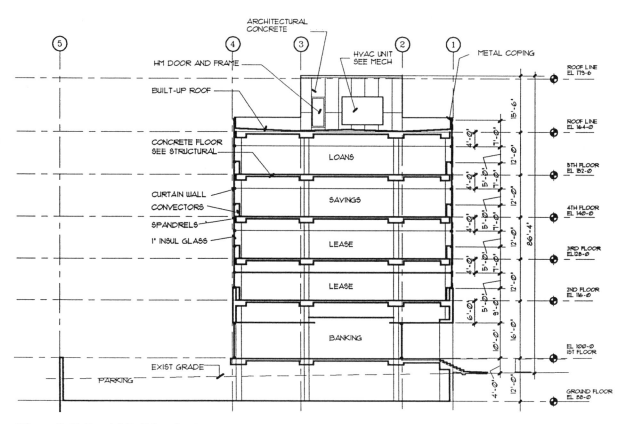

Figure 8.47 *Partial Building Section.*

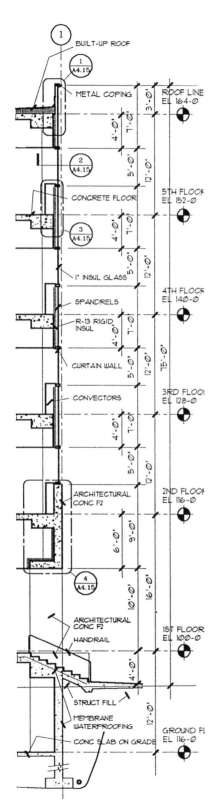

Wall Sections

The wall sections drawn at design development were only the beginning. At contract documents, every exterior wall condition needs its own section view. Normally footings are not shown; let the structural engineer do this. Architecturally, you do need to illustrate perimeter drainage, but do not locate the bottom of footing any specific distance from the lowest floor line. Use cut lines to illustrate a variable distance between footing and floor line. Show drainage devices and note kind in same terms as specified. Show method of damp-proofing foundation walls and waterproofing basement walls. Above grade, show all exterior wall materials and note finishes. Show all flashing. Detail all joints in the exterior of a building. Don't rely on sealant alone. Some day, I feel certain, we'll see that many of the buildings of the 1970s and 1980s will need major repairs to the structural frame because inadequate steps were taken to keep water out. For example, brick or stone veneer applied over metal studs without an exterior sheathing and moisture barrier is a condition asking for problems when water gets into the building. Show structural frame, and proper fireproofing. Draw sections through windows, storefronts, curtain walls, and all conditions of solid wall. Show roofing conditions and detail. Show conditions against existing structures and detail the joints. Show floor lines and vertical dimensioning (see Chap. 2, "Dimensioning"). Show column lines and building lines and dimension recessed facades to the building line. Wall sections can turn into plan sections to show special conditions like rooftop gardens and skylights. Show building insulation and interior materials. Note all materials and reference all details.

Figure 8.48 Wall Sections.

Interior Elevations

The design-development package should have shown most of the required elevations. (Do not draw every wall in a building unless it has something on it that needs elevating.) Interior elevations need to be completed for pricing and construction. Recheck casework and floor plans for accuracy. Check countertops for knee spaces. Add supports where needed. Locate scribes where cabinets abut walls and ceilings and at corner units. Do not assume any building surface is plumb or square. Locate critical items inside cabinets like electrical outlets and communications boxes. Show all deck-mounted sinks and fittings, and coordinate with casework specification. Have them supplied and installed by the casework contractor with hook-up by plumbers. This helps avoid the problem of a sink being too big to fit in a cabinet. For standard casework construction, specify the construction detail by making reference to the standard. For custom casework, detail all edges and joints. Detail wall attachment of all casework and as required by local codes. Show wall-mounted equipment and detail. Check with the structural engineer if weights or moments are a problem. Show millwork such as chair rails and crown moldings. Detail attachment of each. Show chalkboards, tack boards, etc., and detail wall reinforcing. Review all conditions shown and detail supports, shapes, joints, etc. (see Chap. 9 for correct referencing).

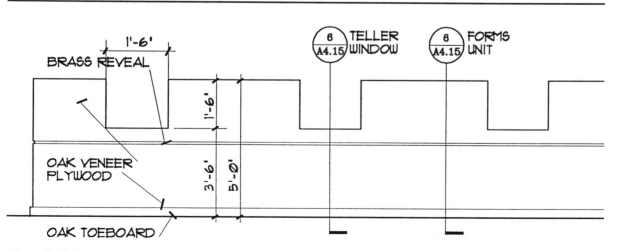

Figure 8.49 Example Interior Elevation.

Reflected Ceiling Plans

The design-development reflected ceiling plans should be substantially complete by this time. Check to add curtain tracks and other ceiling-mounted items or equipment. Detail the supports for these items. In hard ceilings, locate and detail access panels. Do not show fire-sprinkler heads. This is outside the scope of most architectural service and will usually get you into trouble. Locate smoke detectors, exit signs, and any expansion joints. Show ceiling height if not already shown on the Room Finish Schedule, and where more than one height exists in a room. Draw details and reference as shown in Chap. 9.

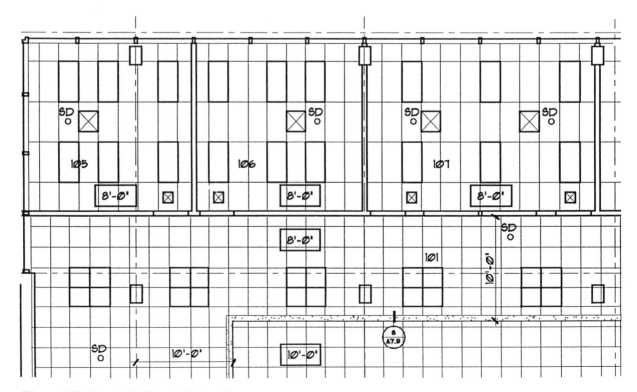

Figure 8.50 *Partial Reflected Ceiling Plan.*

Vertical Circulation

Take the design-development plans and sections and develop the details. Detail stair typical riser and tread showing the nosing for each kind of stair (metal or concrete or wood). Detail ornamental railings, banisters, etc. (see Chap. 2, "Dimensioning," for stair dimensioning). At elevators, double check clearances, especially at the pit and penthouse. Do not allow other equipment in penthouse machine rooms. Check the local elevator codes thoroughly. Some mechanical shafts require special attention due to security or sanitary requirements. Draw plans and sections, and detail finishes and security gratings as needed. For high-rise buildings, the stair and elevator shafts may require pressurization. If this is the case, find room for the fan and the exhaust louvers. Consult with the mechanical engineer for the proper method of pressurization. Exit stairs in a high-rise are often design/build. All that is needed by the architect is a rise-and-run drawing sporting the dimension points of the stair and landings. Work with the stair fabrication to provide adequate anchor locations and in the proper places.

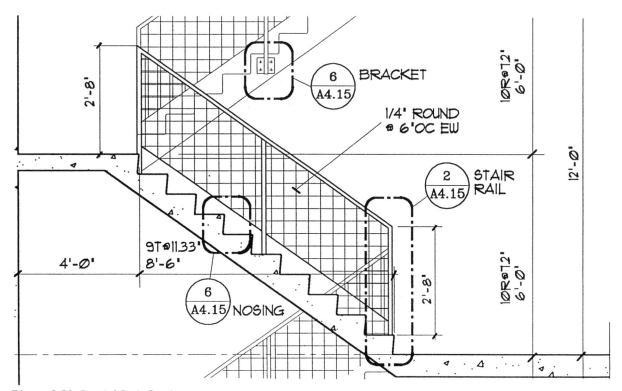

Figure 8.51 *Partial Stair Section.*

Schedule

All schedules were started at design development. Now add detail references. Detail finish conditions at floor transitions, floor and base conditions, wainscot conditions, and ceiling and soffit conditions. (See Chap. 7, "Document Order and Identification" for detail numbering system.) Detail all door, relite, and other opening frames and reference in the Door and Opening Schedule. Dimension all frames. (See Chap. 3, "Door and Opening Schedule.") Detail windows and storefronts and reference them on the Window Schedule. Detail equipment supports and note the detail on the Equipment Schedule. Add any new items to the schedules and complete all blank spaces. See Chap. 3, "Schedules," for individual schedule requirements and complete the work.

Details

The contract document phase is the time when most details are drawn. (See Chap. 2, "Anatomy of a Drawing.") Develop every building system completely. Use the detail book index and the specification-system index to identify places where details are needed. Also print plans, elevations, and sections and "bug" areas that need details. As a criterion, look for elements which define shape, protect the building from water infiltration, and attach something to something else.

DETAILS

REFERENCE LINES

DIMENSIONS

MATERIALS

SIZES

FINISHES

WATER INTEGRETY

CONNECTIONS

Figure 8.52 *Details must show these key elements.*

Interior Design

If interior design work has not started, don't wait any longer. Many elements of interior design ripple down to effect other building elements. Follow the concept for production described under design development. Determine who is to draw final documents and details. Depending on contractual relationships, the architect may draw up the designs of the interiors people. Finish interiors items during contract documents. Do not put off paint and carpet selection, for example; this leads to the selection of materials or colors that are not in the budget or not in stock.

Consultant's Documents

Drawings prepared by mechanical, plumbing, power, lighting, structural, and civil engineers must now be complete. Do not accept the idea that "We can work it out in the field." This will cost you money and maybe even clients in the long run. Review their documents early enough to allow them to respond.

CONCLUSION

During the contract-documents phase, the architect must take the previous drawing efforts and turn them into documents from which a project can be built. From these documents, the contractor must be able to fix a cost and the owner must be able to expect a result. The details must reflect both good design and good technical problem solving. The resulting building must stand for the life intended. If it does not survive with proper maintenance, the architect could be subject to restitution if it can be proved that the design is faulty. A knowledgeable architect can usually avert the conditions that cause building failures through the creation of thorough and accurate contract documents.

CONSTRUCTION PHASE

DRAWINGS ARE LEGAL DOCUMENTS

When the documents are published at the conclusion of contract documents, they are legal and binding on all signing parties. Changes are permitted only through a prescribed format.

First, it must be understood that a numbered change or revision applies to an individual page of a document, not the document as a whole. Further, each revision is dated because it is a change to a contract, and as such needs a legal date of execution.

Every change to a contract document must be made to the original. This is the single-most frequently violated tenet in contract-document administration and one of the most damaging. In fact, please allow me to say it again: *Every change to a contract document must be made to the original.*

Case Study

Consider the actual case of an architect who was designing the retail storefronts for a major downtown building. The drawings went out at the end of contract documents, but they were only about seventy-five percent complete. The elevation views were finished, but few details had been drawn. The first revision was made at the contractor's request for more information. The architect made a paper sepia of the original, and proceeded to revise the sepia. This drawing was then published with the same sheet number as the original drawing. Since the original was so poorly drawn in the first place, three subsequent requests for clarifications were made by the contractor. Each was prepared on a new paper sepia made from the original drawing.

By this time, there was one mylar original and four paper sepia originals, each showing a portion of what was assumed to be a whole picture. No one could tell without a great deal of trouble, or possibly luck, what the whole picture was. So, the contractor asked for a fifth clarification drawing. This time it was prepared on the original, taking revisions 1, 2, 3, and 4 in turn and redrawing them where they should have been in the first place. The resulting combination of revisions showed that conflicts existed, and many other questions were still not resolved. The problem was not solved until the storefront was completely redesigned. The entire effort resulted in a net loss of three months of construction time.

Granted, this was an extreme example, but, by no means the only one where changes were made to copies rather than to originals. Every time this happens, the architect risks working with old information when making decisions. In the example above, had the changes been made to the original document, chances are that the contractor's problems could have been solved after revision number two and maybe two weeks' time. The project would have realized a substantial savings in both time and money.

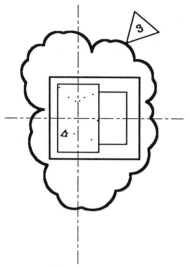

Figure 8.53 *Identify revised area of drawings.*

THE DRAWING REVISION PROCESS

The revision process begins by making a paper sepia copy of the original and filing it as a record of the design. (Never draw on it.) Then, as changes occur, they can be made to the original. When a change is complete, make another paper sepia and file it on top of the previous record sepia. In this manner, the original drawings are always up-to-date and there is a paper trail of every revision made to the drawing which can be retrieved and copied as needed.

Make a Record Copy

Before beginning any revisions, look in the revised-drawing file to see if a sepia record has been made for the last revision. If not, make one as noted above and file it. If this is the first revision, make a sepia to record the original condition and file it. Then prepare the revision on the original drawing. When the revision has been made, it must be identified and issued.

Identify Changes to Existing Drawings

Identify the revision by clouding the area that was changed. Place the revision-indicator symbol so that it points to the clouded area and in the general direction of the change. Enter the next consecutive revision number in the indicator symbol. When more than one revision is made at one time, place the same revision number in each revision-indicator symbol. This is because all changes made and issued to a particular document on any given date comprise the same revision to that document. If another change is made and issued to the same document on the following day, it must be identified by the next consecutive revision number.

Identify Revisions by Adding New Drawings

Revisions to a drawing sheet can also be in the form of a new drawing. For example, drawing sheet A7-15 might contain window details up to number 16. (See "Drawing Sheet," Chap. 1 and "Systems Drafting," Chap. 6.) The next detail drawn on this sheet will be 17. If a new detail is needed, draw it here. Number it 17 and cloud the whole detail, including the title, and identify it with the next consecutive revision number for sheet A7-15. Make record sepias as described above, copy and issue the revision.

Issuing Revisions

Revisions are issued in one of two ways: 8½-×-11 sheets and full-size drawing sheets. In the case of the added window detail above, it would be excessive to reprint the entire drawing sheet just to issue one detail. In this case, copy the new detail including its title and number, and the revision cloud and indicator. Cut it out and paste it on an 8½-×-11 standard drawing sheet. Number the drawing sheet R1/A7-15 if this is revision number 1 or R2/A7-15 if it is revision number 2, and so on. Make the required copies and issue by letter of transmittal.

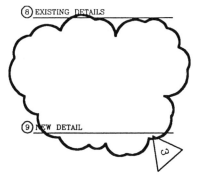

Figure 8.54 *Revision by addition.*

Figure 8.55 *Issue on standard sheets.*

Multiple Revisions to a Single Drawing

This same process is used when a small number of revisions is made to a drawing sheet, but not all of them will fit on one $8\frac{1}{2}$-x-11 sheet for issuing. For example, a plan drawn on sheet A2-6 has three dimensional corrections to be made. The old dimensions are erased and replaced by the correct dimensions. They are clouded and indicated by the revision symbol. Each revision symbol receives the same number which is the next consecutive number for sheet A2-6.

Issuing the Revision

Make a copy of drawing sheet A2-6. Cut out each revision, making sure there is enough of the surrounding drawing to identify its location. Paste these cut-outs on $8\frac{1}{2}$-x-11 standard detail sheets and number each as follows: R2.1/A2.6, R2.2/A2.6, and R2.3/A2.6. It doesn't matter which $8\frac{1}{2}$-x-11 sheet is numbered first, only that they are consecutively numbered following the revision number, "2," in this case. Copy and issue by letter of transmittal.

In both examples above, the revision must be recorded on the sheet in the revision box. Place the next consecutive revision number in the revision symbol, and enter the date of issue next to it.

When a large (more than six) number of revisions is made to a sheet at one time, the entire sheet is reissued. Cloud each revision and identify with the revision symbol as before. Enter the revision number and date in the title block and issue the whole sheet.

REVISIONS TO DETAIL BOOK

When details are published in book form, they are revised in the same manner as described for large-sheet drawings except the number in the detail sheet does not change. The revision number is indicated in the revision box and dated. *Notice:* This revision box is only used if the original drawing was issued in the 8½-x-11 format. If this 8½-x-11 sheet is used to issue a revision on a full-size sheet, this drawing revision area is ignored.

Whenever an entire drawing sheet is added or deleted, be sure to revise the Index to Drawings, and issue it along with the addition or deletion.

Figure 8.56 *Revision to detail book.*

REVISIONS TO SPECIFICATIONS

The specifications are amended in three basic ways. These are additions, deletions, and changes.

Additions are underlined to denote they are new. A revision number is entered in the right-hand margin, at the end of the line, to note the consecutive revision to each specification section.

Deletions are crossed out by overstriking a "/" or a "-" over the data to be removed. Never remove the original text. A revision number is entered in the right margin as before. In this manner, everyone can see exactly what is being removed, and it will stay there as a record for the life of the project.

Changes are entered as a combination of an addition and a deletion. The new material is underlined and the old material has an overstrike. Enter the revision number in the right margin.

```
EDMONDS - HAZELWOOD                              SECTION 09660
ELEMENTARY SCHOOL                        RESILIENT TILE FLOORING

                            SECTION 09660
                        RESILIENT TILE FLOORING


PART 1 - GENERAL

1.01  SECTION INCLUDES

   A.  Work includes but is not limited to following:
       1.  Vinyl composition tile laid over concrete substrate.
           a.  Colors and patterrns as indicated.
           b.  Provided by one manufacturer.

1.02  RELATED SECTIONS

   A.  Coordinate related work specified in other parts of the
       Project Manual, including but not limited to following:

           Section 03300   -   Cast-in-Place Concrete
           Section 09651   -   Cementitious Underlayment:  For
                               leveling floor surfaces
           Section 09678   -   Resilient Base and Accessories
           Section 12300   -   Manufactured Casework
           Section 12000   -   Plastic Faced Casework

1.03  REFERENCE STANDARDS

   A.  Comply with the requirements of Section 01091 and as listed
       herein.  See Section 01091 for listed association, council,
       institute, society, and the like organization for its full
       name and address:

       American Society for Testing and Materials (ASTM):
       ASTM E84-87 Test Method for Surface Burning Characteristics
                   of Building Materials.
       ASTM E648-86    Test Method for Critical Radiant Flux of
                       Floor Covering Systems Using a Radiant Heat
                       Energy Source.
       ASTM E662-83    Test Method for Specific Optical Density of
                       Smoke Generated by Solid Materials.
```

Figure 8.57 Specification change.

Issuing the Change

For small changes that do not affect page numbering, issue the whole page with the revision on it.

For revisions which affect more than one page, there are two routes to take. First, if each new page has the same numbers as the original pages they replace, simply reissue the affected pages. Second, if the new pages are not the same as the original pages, for example, the revision takes three pages to delete and add material shown on two original pages, then number the new pages with the two original pages and add, -a, -b, -c for every additional page.

For revisions that delete information, reissue the pages affected with their original numbers.

Along with each revision, issue the table of contents with any changes in page numbers.

CONCLUSION

Architectural contract documents vary a great deal through the life of the project. A floor plan begun at schematics may survive to become a composite floor plan or code plan at completion of contract documents. A clear understanding of drawing content by phase of work can result in greater efficiency in document production and fewer omissions in drawing content.

FINDING YOUR WAY THROUGH THE DRAWINGS

9

INTRODUCTION

It should take no longer than 30 seconds to find any drawing within a set of contract documents. Unfortunately, this is all too often not the case. In almost any office you will see people spending far too much time leafing through drawings, annoyed because they can't find the particular detail they need. That time spent searching for details or other drawings could, without fail, be put to more profitable use. Developing a good system for finding your way through drawings is what this chapter is about.

The ability to find drawings starts with the drawing order described in Chap. 7. It is a logical order which, when memorized, can help an individual turn to any section quickly. But this alone would still not work when looking for a specific drawing. For that, there needs to be a system for labeling and numbering drawings which works quickly, say within three or four turns of the drawing pages.

Looking for a Window Detail

Let's assume that you are looking for a detail for the windows on the west elevation of a building. You would start at the cover sheet which shows the Index to Drawings. The index says that the west elevation is on sheet A3.5. Turning to A3.5, you see on the west elevation that the window you are looking for is type "C." Turning back to the index, you'll find that the Window Schedule is located on Sheet A5.21. You now turn to sheet A5.21 and find window type "C." On it is identified detail 7/A5.23 as being the detail you are looking for. You turn to sheet A5.23, and there it is.

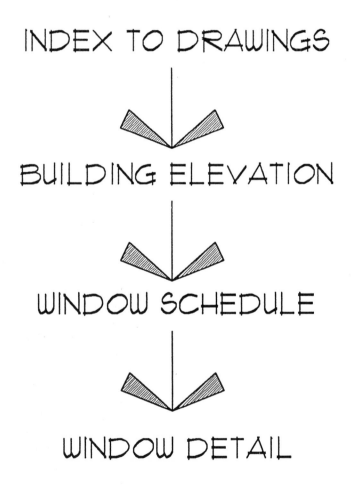

Figure 9.01 *Finding a window detail starts with the Index to Drawings.*

Looking for a Finish Detail

Here is another example: You are looking for a detail of a floor transition that occurs under a door between ceramic tile and vinyl-composition tile floors. Again you start with the Index to Drawings on the cover sheet. It says that the Room Finish Schedule and General Notes are found on sheet A10.1. It also notes typical finish details on sheet A10.2. Turning to sheet A10.1, and reading the General Notes, you find note number 5. Note number 5 says, "See detail 9/A10.2 for typical transition between CT and VCT." Turning to sheet A10.2, detail number 9 is exactly what you were looking for.

Reference in One Direction Only

The philosophy that makes this work is basic. Always reference from the general to the specific in a one-way linear manner. For example, don't try to reference a window type from the head detail. The head detail could work for many window types in the project, so this would not yield a definitive result.

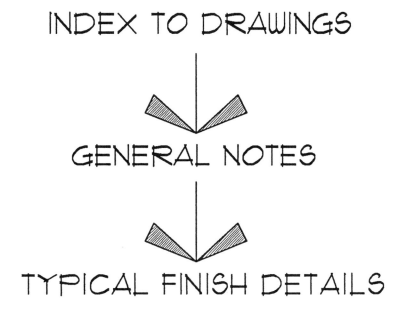

Figure 9.02 *Notes can also reference drawings.*

A Recap of Basics

To begin at the beginning: Every project has the following documents:

Drawings
Schedules
General Notes
Specifications

In Chap. 7, you will find the drawings arranged in the following order:

CHAPTER	TITLE
A0	General
A1	Site Conditions
A2	Floor Plans, Roof Plans
A3	Building Elevations and Sections
A4	Wall Sections, Panels
A5	Building Details and Schedules
A6	Vertical Circulation
A7	Detail Plans
A8	Interior Elevations
A9	Reflected Ceiling Plans
A10	Interior Details and Schedules

Also in Chap. 7 is the system for identifying a drawing by the following:

Phase of work, i.e., SD, DD, CD
Design discipline, i.e., Architectural, Structural

In Chap. 1 are graphic standards for identifying the following:

Sheet title by name and number
Drawing title by name, number, and scale

A Three-Level Referencing System

All information in a set of drawings is found on or through three levels of referencing: *the Index to Drawings*, *primary reference drawings*, or *secondary reference drawings*.

Drawing Index

Start your information search with the Index to Drawings on the cover sheet. This index is not only a list of page numbers. If time is taken to prepare it properly, it is also a useful tool in finding information. (See Chap. 3, "Schedules.") The drawing title is always listed exactly as it appears on the individual drawing sheet including abbreviations and punctuation.

INDEX TO DRAWINGS

AØ	SITE PLAN
A1	1ST FLOOR PLAN
A2	2ND FLOOR PLAN
A3	3RD FLOOR PLAN
A4	ROOF PLAN
A5	BUILDING ELEVATIONS
A6	BUILDING SECTIONS
A7	WALL SECTIONS
A8	EXTERIOR DETAILS
A9	ROOF DETAILS
A1Ø	WINDOW CONDITIONS
A11	OPENING SCHEDULE
A12	DOOR DETAILS

Figure 9.03 *Sample Index to Drawings.*

Primary Reference Group

PRIMARY GROUP
_____

SITE PLAN

FLOOR PLANS

BUILDING
ELEVATIONS
& SECTIONS

CEILING PLANS

GENERAL NOTES

Figure 9.04 Primary reference group.

These drawings are the beginnings of a series of drawings that leads to the construction details. They are the following:

> Site Plan
> Basic Floor Plans and Roof Plan
> Building Elevations and Sections
> Reflected Ceiling Plans
> General Notes

These root drawings, as a group, define the basic scope of the work. Everything else is detail. Further, it is from these root drawings and general notes that all other drawings are found. This is, therefore, the primary location to start a drawing search.

Secondary Reference Group

The second level of drawing in the referencing chain are:

> Schedules
> Wall Sections
> Detail Plans
> Interior Elevations

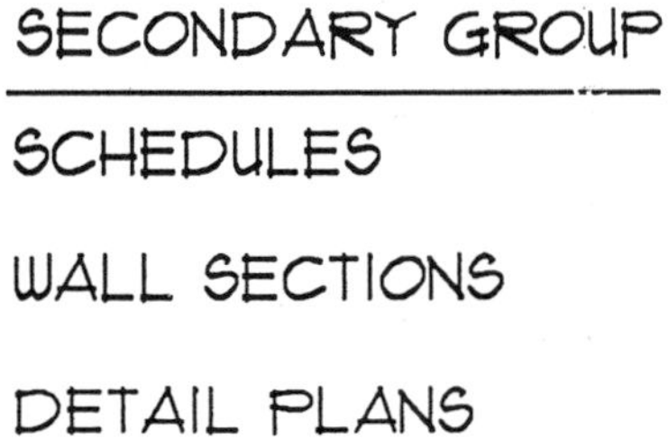

SECONDARY GROUP
_____

SCHEDULES

WALL SECTIONS

DETAIL PLANS

Figure 9.05 Secondary reference group.

Examples

It is from these drawings that all other information is found. Here are a few examples:

> *Example 1:* You are looking for the Level 4 roof plan. It will be noted on the Index to Drawings.

> *Example 2:* You are looking for a wall section through the north wall of the building. The Index to Drawings will list A2.1 as the floor-plan drawing. From there, you will find the section indicator symbol calling out the north wall section for which you are looking.

This system of linear referencing should be used for ninety-five percent of the drawing referencing. Like any rule, however, there are exceptions. Assume that you are detailing an equipment mounting plate and anchors. The Equipment Schedule has made reference to one detail, usually the full-scope detail. Additional cuts may be referenced from this one original detail in order to keep the process smooth and readable. It is not otherwise advisable to cross reference from detail to detail.

REFERENCING SYMBOLS

The following is a quick listing of graphic symbols and titles shown in Chap. 1 and used to reference our way through a set of drawings.

MATCH LINE   Use on composite floor plans to indicate the boundaries and sheet numbers of large-scale plans. Use on large-scale plans to call out neighboring plans without returning to the composite plan.

KEY PLAN AND NORTH ARROW   Use on all large-scale plans which do not show the entire building. Use on building elevation and building section sheets to key the location of the elevation or section. The key plan is not the primary reference for building elevations and sections, only a device used to help the reader see where the view is "cut" without returning to the floor plans.

ROOM NAME AND ROOM NUMBER   Used on floor plans as the primary reference for finishes. They also appear on large scale plans and reflected ceiling plans to identify the space. They also head the Room Finish Schedule.

SHAFT NUMBERS   Appear on floor plans as a primary name for vertical spaces. Also appear in the drawing title for detail plans and sections through these shafts.

FINISHES   When the Schedule of Finishes is used, they are referenced directly on the floor plan under the room title.

DOOR AND RELITE NUMBERS   Are shown on the basic plans and on the Opening Symbol.

BUILDING ELEVATIONS SECTION INDICATORS   Are shown on the basic plan.

WALL-SECTION INDICATORS   Are shown on the basic plan as the primary source. Some people like to show them on building elevations, but this should only be done for clarity, not as a primary location.

DETAIL-PLAN INDICATORS   On basic plans as the primary source.

STAIR-AND-ELEVATOR-SECTION INDICATORS   Show on the larger scale circulation plans as the primary source.

KEY PLAN SECTION INDICATORS   A secondary system for referencing Building Elevations, Building Sections, Wall Sections, and sections through stairs and elevators.

INTERIOR-ELEVATION INDICATOR   Show on the basic plan if no large-scale plans are prepared. Use large-scale plans as the primary location whenever possible.

CEILING-HEIGHT INDICATOR   This notes the height of the ceiling above the reference floor and is located with the room number on the Reflected Ceiling Plan.

WINDOW-TYPE INDICATOR   Shown on building elevations as the primary source. It may be repeated on large-scale exterior elevations.

CASEWORK INDICATOR   When the casework schedule is used, casework items are referenced on interior elevations as the primary source.

EQUIPMENT INDICATOR   Use this on detail floor plans and interior elevations as coprimary sources.

DETAIL INDICATORS   May be used on any drawing to reference a detail.

REVISION INDICATOR   May be used on any drawing to denote a change.

Following is a chart which abbreviates the preceding information for quick reference. The remainder of this chapter will illustrate conditions of drawing referencing.

DRAWING REFERENCING GUIDE

● PRIMARY REFERENCE
○ SECONDARY REFERENCE
* WHEN NEEDED

REFERENCE	DRAWING INDEX	SITE PLAN	FLOOR PLAN	BUILDING ELEV & SECT	WALL SECTIONS	DETAIL PLANS	INTERIOR ELEVATIONS	CEILING PLANS	STAIR & ELEVATOR PLANS	EXTERIOR DETAILS	INTERIOR DETAILS	KEY PLAN	SCHEDULES
DRAWING TITLE	●	○	○	○	○	○	○	○	○	○	○	○	○
DATUM or BENCH MARK		●											
GRID LINES		*	●	○	○	○	○	○	○	○	○	*	
ELEVATION MARKS				●	○		*			○	○		
MATCH LINE		*	●	*									
KEY PLAN & NORTH ARROW		○	●	○	○	○		○	○				
FLOOR PLAN	●												
ROOM NAME			●	*	*	○	○	○					○
ROOM NUMBER			●					○	○				
SHAFT NUMBER			●	*					○				
FINISHES			●										○
DOOR & RELITE			●										
WALL & PARTITION			●										
INTERIOR ELEVATION			*				●						
CEILING PLAN	●												
CEILING HEIGHT								●					○
DETAIL PLAN			●								○		
BUILDING ELEVATION	○		●								○		
BUILDING SECTION	○		●								○		
WALL SECTION	○		●		*			○			○		
STAIR & ELEVATOR SECTION	○		*						●				
WINDOW TYPE	○			●									
LOUVER TYPE	○			●									
WALL PANEL TYPE	○			●									
CASEWORK CONFIGURATION	○						●						
EQUIPMENT						●	●						
DETAILS			●	●	●	●	●	●	●	○	○		●
REVISIONS	●	●	●	●	●	●	●	●	●	●	●	●	●

Figure 9.06 Drawing Referencing Chart.

KEY PLAN AND NORTH ARROW

The primary location for the *Key Plan* and *north arrow* is the title block. It is used on plan sheets when the whole building does not fit on one drawing sheet. The Key Plan is shaded to represent the area of the building drawing on the sheet. It is used on building elevations and building sections to serve as a quick reference for the drawings shown on the sheet. It is also used on wall-section sheets as a quick reference to where the wall sections are taken. And it is used on vertical circulation sheets. Here it is shaded to represent the stair or elevator location in the building.

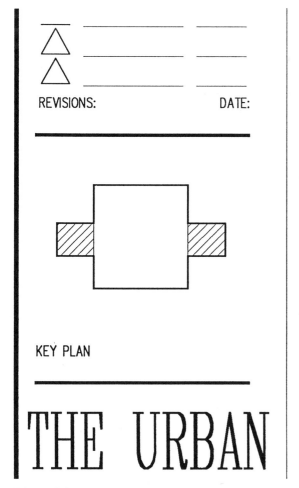

Figure 9.07 Key Plan indicates area of building plan.

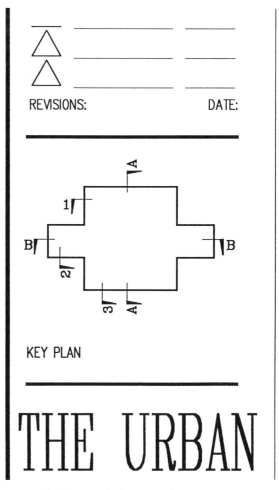

Figure 9.08 Key Plan used to show section cuts.

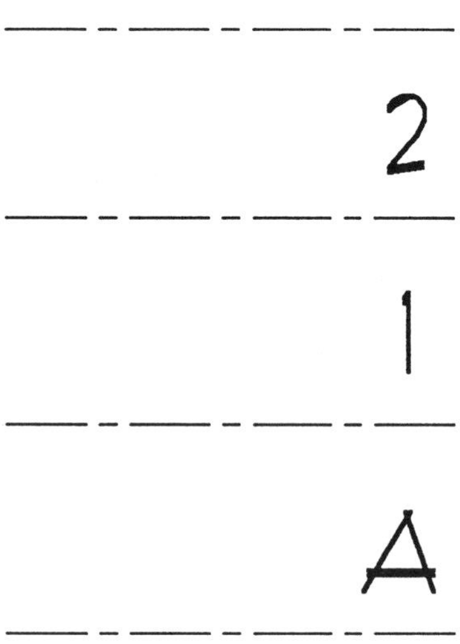

Figure 9.09 *Floor numbering.*

ROOM NAME AND NUMBER

Identify Floor Number

Room numbering begins with a character which represents a floor or level within the building. Floor levels are typically numbered starting at the ground floor with number 1. Each successive floor is numbered consecutively. Basements are often lettered A, B, C, instead of numbered, to denote the fact that they are below grade. Problems arise in buildings which occupy sloped sites. The first floor should always be the floor which opens to the lowest level of grade. In this way, subsequent floors numbered 2, 3, 4, etc. may also have access to grade but are obviously not the lowest level. This is important in large campus-like buildings so visitors will always know which floor they are on when going from one building to another.

Which Spaces Need a Room Number

The room name and number appear primarily on the basic floor plan. It is the address of all interior spaces including:

Rooms
Corridors
Stairs
Shafts (those which have finish requirements)

The room number provides the basis for a variety of schedules, including the following:

Room Finish Schedule
Door and Opening Schedule
Equipment Schedule
Furnishing-and-Accessories Schedule

ROOM NUMBERS

ROOMS
CORRIDORS
ALCOVES
LOBBIES
VESTIBULES
CLOSETS
STAIRS
FINISHED SHAFTS

Figure 9.10 *These spaces need room numbers.*

The architect must take care in inventing a room-numbering system. Here are some criteria to consider:

Room numbers must be unique. Every space in the building must have its own number.

The room-numbering system must also be recognizable by the drawing reader. Many systems are too complicated to follow. The reader must be able to find a room if he or she is given only the room number and a set of drawings.

Shaft Numbering

Shafts are typically numbered with an independent system because they occur on more than one floor. For example:

Elevator No. 1
Dumbwaiter No. 3
Cart lift No. 2
Chute No. 1
Mechanical Shaft No. 2
Stair No. 7

These spaces require room numbers, too, when you want to talk about a particular floor or level within the shaft. This happens with stairs and sometimes mechanical shafts when they need to be finished on the inside.

Three-Part Room Numbering

A simple room number has three parts:

Number or letter to designate the floor or level of the building. Use numbers above grade and use letters below grade.

Numbers for each room on the floor.

Letter suffix may be used to extend a room number into a closet or to identify individual rooms in a suite.

This simple system can be expanded to accommodate the numbering of rooms in a building with wings or extending elements. In this case, add a letter prefix which represents the wing or area of building.

Start numbering the building at the front entrance. This space will be number 100. Number lobbies, vestibules, corridors, and other circulation areas with a letter suffix such as 100A, 100B, 100C, etc. This way, all circulation spaces are easily recognizable as a 100-series number. Number spaces consecutively. Add letter suffixes to denote rooms in a suite. Add letter prefaces to designate areas of the building.

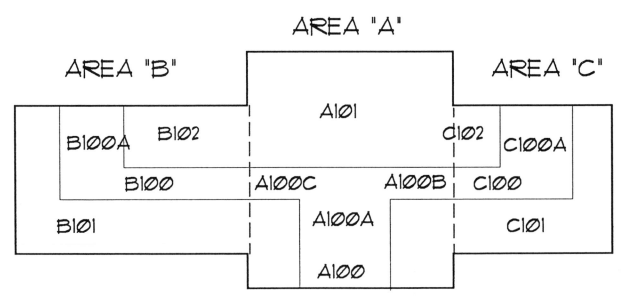

Figure 9.11 Partial plan with room numbers.

FLOOR PLAN AND ROOF PLAN REFERENCING

The following referencing symbols are primarily used on the floor plan:

Detail-Plan Indicator
Match Lines
Building-Section Indicators
Building-Elevation Indicators
Wall-Section Indicators
Room Name and Number
Shaft Name and Number
Room Finishes when Shown on Schedule of Finishes
Door and Relite Numbers
Partition Indicators
Detail Indicators
Equipment Indicators
Revision Indicators

Each of these symbols leads directly to another drawing with information that could not have been shown on the plan drawing. In some cases, for example, the Schedule of Finishes, this is the final destination in a search for information. In other cases, for example, the wall-section indicator, this is the first step in finding a wall-flashing detail.

See the sample floor plan for the use of each symbol. Notice the reference to a column detail.

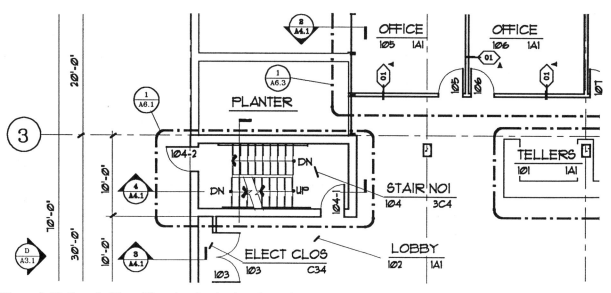

Figure 9.12 Sample Floor Plan showing basic referencing.

BUILDING ELEVATION AND BUILDING SECTION REFERENCING

The building elevation has its own list of referencing symbols. These include the following:

Window and Louver Indicators
Match Lines
Building Section Indicators (not the primary location)
Wall Section Indicators (not the primary location)
Exterior Detail Indicators
Detail Indicators
Revision Indicators

Each of these symbols references another drawing or schedule where additional information can be found. The sample building elevation shows the use of each symbol (except match line) and an explanation of how to get to a window detail.

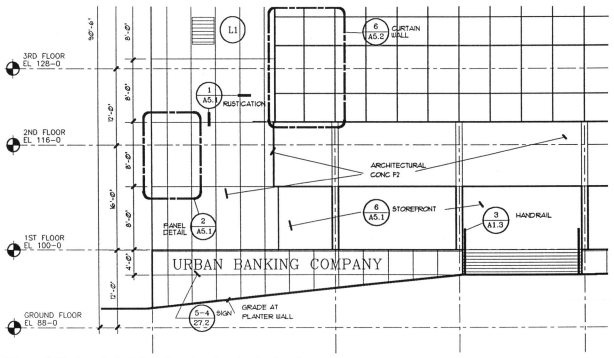

Figure 9.13 Sample building elevation showing basic referencing.

Wall Section Referencing

Wall sections are found by reading the reference symbols on the floor plans, and if used, on the building elevations. From the wall section, a tertiary level of information can be found. The symbols used for referencing are:

Detail Indicator
Revision Indicator

The detail indicator leads the reader to every detail condition shown along the length of the section, from footing drainage to window details to roofing details. Some of these details will also be shown on the Window Schedule. The roofing detail will also be repeated on the Roof Plan.

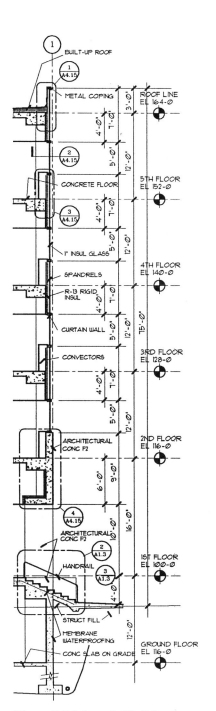

Figure 9.14 *Sample Wall Section.*

REFLECTED CEILING PLAN REFERENCING

Reflected Ceiling Plans deal with the information on the ceiling. Therefore, referencing symbols deal mostly with this area. They are as follows:

> Room Numbers
> Detail Indicators
> Revision Indicators

It is not customary to show wall-section indicators on the ceiling plan, but if they help to clarify the intent, it can be done. Most of the referencing made on the Ceiling Plan is in the form of detail indicators used to find details for edge conditions and soffits. See the sample Ceiling Plan showing the use of these symbols and the path to a soffit detail.

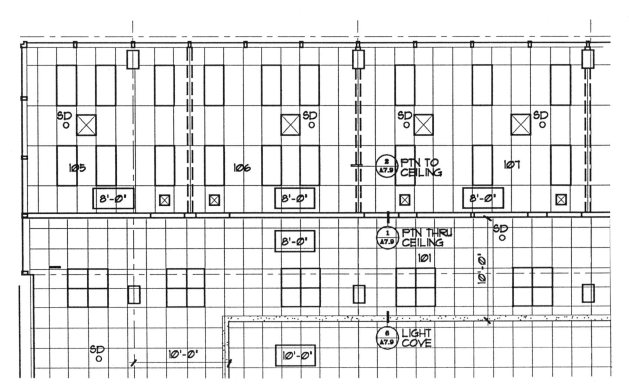

Figure 9.15 Sample Reflected Ceiling Plan.

SITE-PLAN REFERENCING

The Site Plan can contain a variety of drawing reference symbols. These include the following:

Detail-Plan Indicator
Section Indicator
Detail Indicator
Revision Indicator

The Site Plan is often drawn at rather small scale so any area with pavers would likely be drawn at larger scale. This jump in drawings is bridged by the detail-plan Indicator. Most other references are detail indicators, pointing directly to the site-work details.

See the sample Site Plan showing the use of each symbol and path to a planter detail.

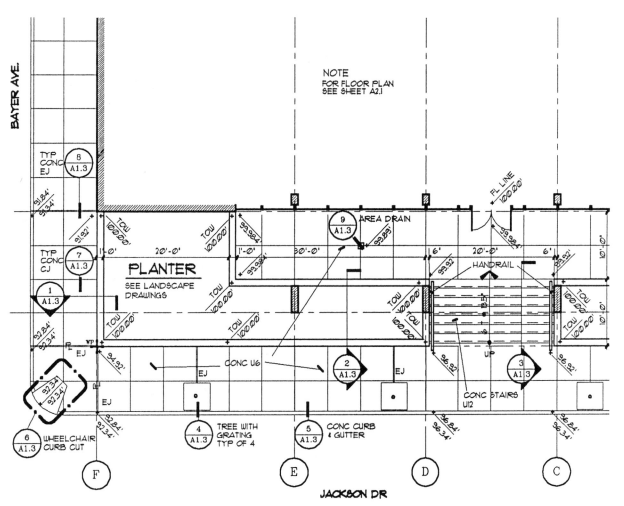

Figure 9.16 *Sample Site Plan.*

CONCLUSION

The system described here is graphically illustrated below. It begins with the Index to Drawings and ends in the construction details. It considers the various paths to each piece of information.

To avoid confusion and to help the reader find information, always reference drawings in the linear fashion shown. Do not try to back-reference.

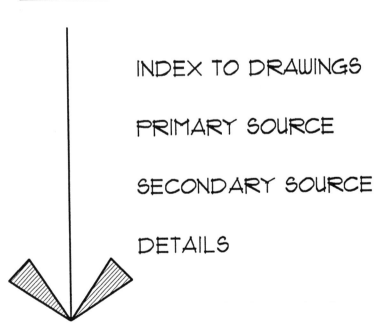

Figure 9.17 The path to drawing referencing.

DOCUMENT QUALITY CONTROL

10

INTRODUCTION

Many architectural projects are too large for one person to design, engineer, and draw alone. Few of us have the knowledge to prepare architectural, interior, structural, civil, mechanical, plumbing, fire-protection, power, lighting, communications, and acoustical drawings without the help of others. Since some of these items are a part of most work, there is a need to hire staff and consultants. With this need comes the problem of coordinating their work.

One of the best tools for promoting consistency and accuracy in contract-document production is having standard methods for doing things. This is what this chapter, and in a larger sense, this whole book, is about. By using consistent and expected methods of doing repeat tasks, the results will more likely be consistent and expected. To that end, each design firm should have both *production standards* and *management standards.*

PROJECT START-UP

Distribute Production Standards

It is the combination of good production and management standards that fosters well-coordinated documents. At the beginning of every project, all design-team members should have a copy of the production and management standards. By spending a few hours becoming acquainted with those manuals, each team member becomes better able to fit into the team and work toward the common goal. Don't forget the owner, who is a team member, too.

Prepare a Responsibility Matrix

Once everyone has had time to review the standard, and before contracts have been signed, call a *start-up meeting*. Determine who is to be responsible for what. AIA documents B161 and B162 are very valuable tools at this meeting. They breakdown a project, by phase, from predesign to construction management.

Each phase lists general and specific tasks on a *matrix* with team-member responsibilities. The form is intended as a tool for determining scope of work and method of payment; however, its role can be expanded to team coordination without direct association to the owner/architect agreement.

If you do a lot of design-build work, it is useful to modify the B162 to include tenants and contractors as team players. Also, modify the list of tasks to include items specifically assigned to people outside the architect's office. Along with the matrix, modify and add to item descriptions, so there is a written record of each task. Some specific items to consider are:

Who is doing the energy analysis?
Who is preparing the equipment list?
Who is coordinating purchase and delivery?
Who is moving existing equipment into the new site?
Who is doing signage and graphics?

There are many more, but this gives you an idea of the kinds of questions which need to be explored.

Once the task/responsibility matrix has been agreed upon, it can become a part of everyone's contract document. If enough time is spent up-front, there will be less room for argument and misunderstanding down the line. You don't want to go for a building permit forgetting the energy calculations, just because no one mentioned it at project start-up.

Keep a copy of AIA documents B161 and B162 on file for every project. Keep your own additions handy, so the lessons learned last time can be applied to the next job.

Prepare a Critical Path

When everyone knows the task breakdown, a schedule must be prepared showing both who is responsible and when the task is due. Sometimes subtasks are necessary, such as a soils report, before the footing design can begin. Prepare a flow chart or *critical path*. Break down all work by phase as shown on AIA document B162. This, then, becomes the road map for producing the project.

Determine Production Systems

Another agenda for project start-up is the *production system*. Chapter 6 deals with various production systems, including cut-and-paste, systems drafting, overlay drafting, CAD, and CAP. The system best suited to the project and the design team must be determined and agreed to by all. Today many architects and engineers use CAD, but there are many varieties out there. If CAD is the chosen system for production, be sure everyone has the same program. If not, the next-best thing is to translate files among dissimilar programs. Before agreeing on CAD, exchange copies of standard layer assignments (see Chap. 6) and symbol libraries, and translate a few files to determine compatibility. This becomes even more important when more than two different CAD systems are being used by the architect, structural, mechanical, and electrical designers. The time to do this coordination is at project start-up.

Many design professionals use manual drafting techniques like overlay drafting. There are methods for integrating these techniques into CAD operations. When this mix of techniques is necessary, contact your reprographics house to discuss methods of photographically combining documents. Make sure everyone understands the physical requirements of the process before starting drawings.

Coordinate Layering

Along with production technique comes the question of who will prepare what part of a drawing. The *layer system* of both overlay drafting and CAD allows various team members to prepare their own portion of a drawing. The architect usually prepares the background building plan. Mechanical and electrical engineers produce their work over this background. The structural engineer may use the architect's structural layer, but usually produces his or her own. I don't agree with this practice, but it happens because computers used by structural engineers are not generally compatible with computer systems used by architects. As a result, structural engineers often need to produce their own drawings. Another group to use the architectural base plan is the interior designer.

With this diverse demand on the architectural background plans, great care must be taken in planning the drawing contents. The best way to do this is to list every drawing on a matrix showing its composition of base and overlays. (See Chap. 6, "Overlay Production.") Note who is responsible for each element and how the final document is to be produced.

Cartoon the Drawings

Cartoon every drawing so all design-team members know which drawing sheets are theirs and which parts are shared by more than one team member. For example, the architectural overlay might use half of the plan sheet for Finish Schedule and Notes, and mechanical engineers will use the same space for their notes and schedules. This could only happen with overlay or CAD drafting, and must be planned for at start-up.

DRAWINGS BY OTHERS

Some drawings are produced by persons employed directly by the owner under separate contract outside the architect's responsibility. Two such examples are land surveyors and equipment vendors. The contents of these drawings must be coordinated with the architect's drawings and those of all consultants.

SURVEY DRAWINGS

The land survey and soils report are prepared by the owner's forces, not the architect. This is done specifically to relieve the architect of the responsibility for its contents. It does not, however, relieve the architect and consulting engineers from correctly interpreting the information shown on those documents. Further, it may not relieve the architect from problems not shown on these documents. It is advisable for the architect to become involved with the owner in determining what information is needed on the survey. To that end a *Survey Data Form* should be created. On it, provide space to fill in the information needed. This form should be filled in by the architect in meetings with the owner, so both parties can agree on the scope of the surveyor's services. It then becomes a part of the contract between the owner and surveyor.

Some of the entries to check on the Survey Data Form are drawing scale, north orientation, and drawing media. These are used to inform the surveyor that you want the final drawing on specific size sheets, with north up or left and at a scale of, say, 1 inch equals 30 feet. You can request the drawing to be produced by CAD and the resulting file to be available for your use. By using this form, both the information needed and the method by which it is produced will be compatible with your production system. It really looks bad when the survey and site plans are drawn at different scales and with north opposing each other, so use the Survey Data Form to help coordinate this production need.

VENDOR DRAWINGS

Another document furnished by the owner actually comes from her or his *equipment vendor.* This is a place where far too little time is usually allocated, and way too much money is usually spent. Equipment vendors, like consulting engineers and architects, all have their own production techniques, and they don't automatically conform one to another. Vendors prepare drawings more akin to shop drawings. Their purpose is to show the installation crew how to place the equipment. It is the architect's responsibility to glean from these drawings the information necessary to make sure the equipment will work within the room or building, and that building systems of structure, mechanical, and power are fully coordinated.

The single-most important element in coordinating vendor-furnished equipment is to get the owner to sign a contract with the vendor early enough to get information when it is needed. In medical work, for example, the owner usually doesn't like to sign for a new X-ray machine until the last possible date. This gives the owner time to shop around and make up his or her mind. It also puts the design process farther and farther behind. (Remember the "critical path" mentioned earlier in this chapter.) Late information from owner/vendors can result in design cost overruns and construction change orders. With this knowledge up-front at project start-up, the owner should be informed of the consequences of late information and the likelihood of additional fees, should that occur.

The second-most important issue is to deal very closely with vendors so that architectural considerations are integrated into their installation drawings and vice versa. Consider the vendor as another team member. Distribute production and management standards as well as project schedule and, specifically, any deadlines affecting the vendor. Determine the need for additional subconsultants such as acoustic engineers and radiation physicists and who will pay for their services. Meet regularly with the vendor in the presence of the owner, so all decisions have the owner's concurrence. Never allow the owners to make changes to their vendor's documents or modify the purchase order without your knowledge.

A useful tool for coordinating vendor-furnished equipment is the *Equipment Data Form* (see Chap. 3 on equipment schedules). This form should be completed in the presence of the equipment vendor, so all information is complete and correct. Complete all spaces, even if catalog cuts and brochures are included which repeat the information. When mechanical, plumbing, and power requirements are discussed, the appropriate consulting engineers or design-build contractor should be present.

JOINT VENTURES

Some projects are either too large for small firms or have a complexity which precludes some firms from attempting to provide service alone, such as a marine laboratory or a multipurpose sports arena. In these cases, a *joint venture* between two firms with unique qualifications may be the only way for either firm to get the job.

When such a relationship is considered, there is an important coordination responsibility for both firms. Usually one of the firms has a position of general responsibility for the project, while the other has a supporting role as lead designer or specialty planner. The lead firm with production responsibility should present their production standards as the project standards and meet with the supporting firm to review the standards and agree on their use. It is paramount to the owner in this relationship that the design joint venture appear and work as one, and the documents presented to the owner are a powerful tool to show such unity or the lack of unity.

One of the first signs of design-team unity is a *customized drawing sheet* giving proper billing to both parties in the joint venture. Given sufficient budget, this process should continue on through letterhead, transmittal forms, and other business forms. This practice shows the owner a commitment by both firms to the joint venture and the project.

Once this unity is established, the fine points of project responsibility should be established, as described at the beginning of this chapter. Both parties need to work hard and with direction to ensure the unity of the joint venture for the duration of the project. With a clear definition of responsibilities up-front, there will be little room for disputes as the project continues. With poor definition of project responsibility, there could be voids and overlaps in the work, resulting in in-fighting and even a dissolution of the joint venture before the project is finished. The joint venture can be a rewarding experience for both parties if everyone knows his or her position and responsibility from the beginning.

WORKING WITH CONSULTANTS

Most projects require at least mechanical and electrical *consultants* because the architect simply doesn't know how to provide their services. Along with the need for consulting service comes the need to integrate their work into the project.

Once project start-up has assigned basic responsibilities and published the schedule, the daily task of coordinating the work begins. One method to ensure team coordination is regularly scheduled progress meetings. Monday 10:00 a.m. is a good time. It gives all those involved time to prepare before they arrive, and sets the pace for the remainder of the week.

The agenda should always start with collecting time sheets from team members. Compare actual time with expected performance, then go on to old business. See that old items are cleared from the list quickly, so they do not become a lingering problem. Next is new business. The reference for each section is the project schedule and project plan. Finally, set goals and responsibilities for the week. Be prepared to take last week's time sheets and adjust your plan for the coming week if need be. Use this meeting to discover problem areas and get them solved before they become overwhelming. Do not allow any team member to consistently put off her or his work. This is common when engineers have too little fee. They don't want to start engineering until they feel the architecture is set, so they don't have to do their work twice. The solution to this problem is to stay focused as an architect, and require work on time.

Consultant Coordination

Require each consultant to send sepia drawings for checking. Then you can return the sepia after copying the comments for yourself. As drawings come in from the consultants, review them carefully. Make sure that they follow the plan cartooned at project start-up. Take their drawings and check for background errors. Next start checking one system, for instance, supply-air ducting, until all items of that system have been checked. Then start another system until the drawing has been checked. Use tools like the Room Data Form and Equipment Data Form (see Chap. 3) to verify that everything is covered. Prepare checklists by discipline to remind you of items that could otherwise be overlooked.

When consultant drawings have been checked, make a blueline copy for your record and for further coordination. You can also copy drawings for other consultants' coordination. Return the sepias to the consultants. Require them to return the sepias with their indication that pick-ups have been made. This method provides a closed loop of checking which starts and ends with the architect.

You will often find the consulting engineers unwilling to return check sets. The main reason is the time it takes to administer this checking process. *A word of caution:* As project complexity increases, the need for this coordination also increases. For your own protection, place this requirement in their contract and enforce the review process.

INTERNAL COORDINATION

Internal coordination begins with production standards. It is much like creating a language. Without it, no one will be understood. Whereas with it, the team can focus on design and technical problem solving. The production standards are one of the most important documents your office will ever have. They save time so that fees can be reduced and profits increased. With production standards, you can prevent some of those errors and omissions which are not only costly but embarrassing.

Begin by preparing abbreviations, terms, and graphic symbols. Add standard drawing-sheet and schedule formats and enforce a uniformity in the appearance of all documents. Develop a tightly knit set of drawings that establishes a good reputation for your firm.

Having and using good production standards is just the beginning of internal-production quality control. Each project will have its own problems, but the systems for solving those problems should be in place so the right questions are asked at the right time. Just like with consultant coordination, develop an architectural checklist for checking the work in progress. Do not leave this to chance or to spot checking. You need a system, something to force you to take the time, and a format to use to do it right. (See the "Architectural Checklist" at the end of this chapter.)

SET UP A QUALITY-CONTROL PROGRAM

Many firms discuss the importance of checking drawings and establishing standards, but often this vital aspect of any office practice goes by the wayside. The importance of establishing a good *quality-control program* can't be stressed enough. While it is unusual, there are clients who now ask a firm to submit a copy of their production standards and quality-control program as a requirement of their RFQ (*Request for Qualifications*). If a firm is sued or goes into arbitration, adherence to a published quality-control program could be a deciding factor in the architect's favor. While the cost of building and maintaining a quality-control program may seem high, the benefits will far outweigh the costs in the long run.

There are two basic approaches to performing quality control. First, dedicate the time of one person as manager of the task, and second, make it a committee responsibility.

If the practice is large enough, a *quality-control position* should be established. Call it Partner-In-Charge of Quality Control, Quality-Control Director, or Quality-Control Manager, anything that places it high in the overall business organization of the company. The position must "sound" like one of authority to the staff at large and to potential clients. This person must then have the authority suggested by the title. The Quality-Control Manager will work with other key persons to establish policy and enforce the standards. The company ownership must back the Quality-Control Director, or the program will fail. Time must be allocated to the task. The director should be a full-time employee doing only quality control. The job is that big. Anything less will result in an unsatisfactory program.

When a practice is too small to employ a full-time Quality-Control Director, a committee of partners and selected staff can run a quality-control program. The committee needs a leader, and not necessarily the boss. A person with exceptional technical and production skills should be selected as the Quality-Control Chairperson. This person needs a percentage of time dedicated to the execution of policy and the day-to-day monitoring of the system. The remaining members should represent the design and management forces of the firm, so that a well-rounded quality program can be established.

Duties of the Quality-Control Director

The Quality-Control Director or the committee needs to have basic duties and responsibilities. While there are many duties which fall into the category of quality control, there are a few which should always be a part of that position or group. Following are some broad-scope responsibilities that come into the scope of quality control:

Establish production standards.
Establish a library of standard forms, schedules, and details.
Maintain a library of past project details.
Maintain a codes library and code-review policy.
Maintain the technical library and sample library.
Assist project teams with technical problem solving.
Establish a master specification.
Establish a continuing education program.
Establish construction-administration policies.
Prepare a budget for achieving these goals.
Act as mentor in the code-review process.
Establish cost-control procedures.

Production Standard

Production standards were discussed at the beginning of this chapter and throughout this book. They are the backbone of the program, and must be maintained and enforced.

Library of Standards

It is recommended that the Library of Standards be organized as a series of three-ring binders divided into the detail categories established in Chap. 7. Each volume should contain details considered "standards" for the firm and be available for use "as is" on selected projects. Examples include a concrete curb and gutter, a hollow-metal door frame, and a plastic laminated countertop. The Library of Standards should also include standard schedule forms (see Chap. 3). Each schedule and instruction for its use should appear in the appropriate volume, i.e., Finish Schedule in vol. 7 and Door and Opening Schedule in vol. 5. The library should also contain General Notes which relate to their particular schedule and detail contents. The Library of Standards is intended as a quick reference for standard items, and is not intended to be used for nonstandard conditions.

Library of Past Project Details

A *library of past project details* can be very useful in solving problems on new work. This library is best filed in job-number order, and the contents of each detail group should be kept together. The job-number order will help aid retrieval, while grouping details as originally produced will help future researchers understand the reasons for the detail contents. When taken out of context, some details could be misleading and cause problems when used on a new job. Entries into this library need careful review to help promote top-quality technical problem solving and not misuse of inappropriate details.

A Technical Library

A *technical library* of manufacturer and vendor product samples and catalogs is a primary source of information in the architect's office. It should always contain the products used most often in the practice, with frequent updates by product representatives. The backbone of this library is the Sweets catalogues published by McGraw-Hill. This quick reference, along with back-up reference service, is ideal for getting started on a research project, and a valuable source when manufacturers catalogs are not present in the library. Keep this library current. Old information in this library can be worse than useless. In unskilled hands, this information could lead to drastic errors and costly change orders during construction.

Library of Codes and Standards

Maintain a *library of codes and standards.* This library should contain the building codes enacted by the state(s) and city(ies) in which you do business. This includes, at least, building, accessibility, life-safety, fire, mechanical, plumbing, and electrical codes. Many agencies provide subscription services, like ICBO (International Conference of Building Officials) and NFPA (National Fire Protection Association) which update their material periodically. These subscriptions should be maintained, so that the latest information is always on file. Some standards to acquire are *American Standards Testing Material* (ASTM), design/trade publications like those published by the American Concrete Institute (ACI) and the Gypsum Association *Wall and Ceiling Manual.*

Member of Each Project Team

The Quality-Control Leader should be considered a source of reference just like a library book. The years of experience and general knowledge of other work in the office are valuable references which should not be overlooked. Project teams can use this reference at any time, not just at quality-control review sessions. The Quality-Control Leader should not be viewed as an outsider, or a nitpicker, but as a team member with valuable contributions.

Specifications

Under the Quality-Control Leader is the *Specifications Writer.* Whether an employee of the firm or a consultant, this person is responsible to the office of quality control and assists in establishing policy.

Among the responsibilities of the Specifications Writer is to prepare master specifications for the firm based on project type and phase of work. This will result in a large database of information that can be used and edited for future work. Keeping this database current is a joint responsibility of the Specifications Writer and the Quality-Control Leader.

Continuing Education

Continuing education is important to the livelihood of every practice. We work in a very fast-changing profession, and keeping up with latest trends, products, and rules is very important. One method for staying on the cutting edge is continuing education. A formal program of events is essential. Anything less than planned bimonthly sessions for a formal presentation is not adequate. These sessions can have a number of formats, including lunchtime vendor and product demonstrations, last-hour-of-the-day critiques of current projects, and field trips to projects in progress. The topic material is endless, and worthy of everyone's participation. Vary the topics to include something for designers, technical architects, and managers alike.

Construction Administration

Within the architect's office, the task of *construction adminis-tration* is generally managed in one of three basic ways. In small practices, the principal is often the person to handle this task. As the office grows, additional staff must be brought in. This is done by hiring a Construction Administrator, or by assigning production staff the responsibility once construction begins. As an office grows, and more people become active in construction administration, there is a need for a point of control and formal policy. A person should be named in charge, and located in the overall department of quality control. The primary duties of this position would be to set guidelines for actions taken in the course of this phase of work, including submittal reviews, pay requests, reasons for rejecting work, and preconstruc-tion procedures.

Code Review

The number of codes regulating the design process is growing in both size and complexity. This creates the potential for important issues to be overlooked. I have heard "war stories" from architects who had their building permits denied because of major code problems. This could have been avoided if greater attention had been given to code analysis at project start-up. Often an individual within the firm has a knack for the codes. This person should be assigned, under the Quality-Control Leader, to review all projects for basic code compliance. The project team should remain the primary code source for land-use decisions and code issues relevant to specialty items, i.e., radiation protection, health care, and laboratories. The code expert should work closely within quality control to disseminate information and establish procedures.

Construction Cost Control

One of the primary responsibilities of the architect is to design within the owner's construction budget. The person in charge of *construction cost control* is responsible for assisting the design team with this task. In small offices, construction cost control may be performed by an outside consultant. Given a work load of adequate size, construction cost control can be done in-house by a cost estimator working under the department of quality control. In addition to estimating construction cost, duties include working with project design teams to investigate alternatives, and to train staff in information gathering relative to costs.

Establish a Budget

The whole quality-program works best when it is properly funded. Those field trips can be conducted on personal time, but if there is a *budget* for the hours and staff can be paid, they will be more likely to attend, and morale will be better. Other items to fund include library reference material, computer software, guest speakers, and all or part of the quality-control salary.

CONCLUSION

All of these programs within the realm of quality control will help the architectural practice produce better and more accurate drawings. Graphics and production will be standardized, coordinated, and have a better chance of being produced within a budget. Not the least of all, production quality control will help reduce errors and omissions. These errors and omissions cannot only negatively impact a firm's reputation, but can be very costly and time-consuming. This feature alone would justify setting up and maintaining such a program.

ARCHITECTURAL DOCUMENT CHECKLIST

CODES AND AGENCIES

Review the requirements of codes and lending organizations, and confirm the results with the proper authority. Some items to consider are the following:

❑ Land Use

❑ Historic Preservation

❑ Building Codes

❑ Fire Department

❑ Accessibility Standards

❑ Engineering/Streets Departments

❑ Energy

❑ _____

❑ Shorelines

❑ Environmental Protection Agency (EPA)

❑ Occupational Safety and Health Administration (OSHA)

❑ Health Department

❑ Utility Regulations

❑ Housing

❑ _____

❑ _____

FLOOR PLANS

Review floor plans for compliance with codes, owner program, and the following:

❑ Demolition of Existing Conditions

❑ Key Plan

❑ Horizontal Dimensional Control

❑ Wall Openings

❑ Equipment

❑ Room-Finish Indicators

❑ Wall/Partitions with Types

❑ Stairs

❑ Casework/Millwork

❑ Plumbing Fixtures and Fittings

❑ Interior-Elevation References

❑ Consultant Interface

❑ Earthquake Control

❑ _____

❑ North Arrow

❑ Material Poché

❑ Grid Lines

❑ Fixtures

❑ Room Names and Numbers

❑ Door Swings and Numbers

❑ Wall-Section References

❑ Elevators

❑ Furnishings

❑ Interior-Elevation References

❑ Detail References

❑ Expansion and Contraction

❑ _____

❑ _____

SITE CONDITIONS

Review the site drawings for compliance with codes, owner program, and the following:

❑ North Arrow

❑ Legal Description

❑ Soils Investigation

❑ Property Line

❑ Easements

❑ Accessibility

❑ Locate Building in X, Y, and Z

❑ Site Drainage

❑ Paving and Parking

❑ Furnishings

❑ Irrigation

❑ _____

❑ Site-Location Map

❑ Survey

❑ Bench Mark

❑ Land-Use Restrictions

❑ Setbacks

❑ Existing Conditions to Remain

❑ Demolition

❑ Erosion Control

❑ Utilities

❑ Landscaping

❑ Consultant Interface

❑ _____

SITE DETAILS

Review the site details for compliance with codes, owner program, and the following:

❑ Demolition of Existing Conditions

❑ Site Improvements

❑ Curbs and Gutters

❑ Fences and Gates

❑ Playground Equipment

❑ Gratings

❑ Retaining Walls

❑ Consultant Interface

❑ Earthquake Control

❑ _____

❑ Buildings

❑ Paving and Surfacing

❑ Materials and Finishes

❑ Bike Racks

❑ Site Furnishings

❑ Railings

❑ Stairs and Ramps

❑ Expansion and Contraction

❑ _____

❑ _____

ROOF PLAN

Review roof plans for compliance with codes, owner program, and the following:

❑ Demolition of Existing Conditions

❑ Marquees

❑ Awnings

❑ Building Projections

❑ Drains and Scuppers

❑ Slopes

❑ Material/Systems

❑ Curbs, Crickets, Saddles

❑ Skylights

❑ Walk Surface

❑ Screens

❑ Detail-Reference Indicators

❑ Consultant Interface

❑ Expansion and Contraction

❑ Wall-Section References

❑ _____

❑ _____

❑ _____

❑ Canopy

❑ Overhangs

❑ Perimeter Parapet Walls

❑ Fascia Line

❑ Splash Blocks

❑ Spot Elevations

❑ Rooftop Equipment

❑ Penetrations

❑ Roof Access

❑ Window-Washing Equipment

❑ Access Ladders

❑ Screen-Reference Indicators

❑ Duct Penetrations and Pipe

❑ Earthquake Control

❑ _____

❑ _____

❑ _____

❑ _____

BUILDING ELEVATIONS

Review building elevations for compliance with codes, owner requirements, and the following:

- ❏ Demolition of Existing Conditions
- ❏ Masonry Coursing
- ❏ Wall Openings
- ❏ Curtain Wall
- ❏ Storefronts
- ❏ Reference Schedules
- ❏ Material and Finish
- ❏ HVAC
- ❏ Chimneys and Stacks
- ❏ Ornamentation
- ❏ Fascia
- ❏ Cornice
- ❏ Gutters
- ❏ Splash Blocks
- ❏ Earthquake Control
- ❏ Awnings
- ❏ Mechanical/Electrical on the Facade
- ❏ Light Fixtures
- ❏ Laundry Vents
- ❏ Wall-Section References
- ❏ _____
- ❏ _____
- ❏ _____

- ❏ Vertical Dimensional Control
- ❏ Floor Lines
- ❏ Windows
- ❏ Louvers
- ❏ Doors
- ❏ Grade at Building
- ❏ Rooftop Elements
- ❏ Flagpoles
- ❏ Antenna
- ❏ Trim
- ❏ Canopies
- ❏ Gables
- ❏ Downspouts
- ❏ Expansion and Contraction
- ❏ Signs
- ❏ Consultant Interface
- ❏ Hose Bibs
- ❏ Fire-Department Connections
- ❏ Detail-Reference Indicators
- ❏ _____
- ❏ _____
- ❏ _____
- ❏ _____

BUILDING SECTIONS

Review building sections for compliance with codes, owner requirements, and the following:

- ❏ Demolition of Existing Conditions
- ❏ Floor Lines
- ❏ Materials and Finishes
- ❏ Interior Elevations Beyond Cutline
- ❏ Schematic of Building Components
- ❏ Ceiling Cavity
- ❏ Grade at Building
- ❏ Penthouses
- ❏ Chimneys and Stacks
- ❏ Ornamentation
- ❏ Fascia
- ❏ Cornices
- ❏ Expansion and Contraction
- ❏ Joints
- ❏ Stair
- ❏ Mechanical
- ❏ Consultant Interface
- ❏ _____
- ❏ _____
- ❏ _____
- ❏ _____

- ❏ Vertical Dimensional Control
- ❏ Building Elevations Projecting from Section View
- ❏ Foundations and Break Lines
- ❏ Room Names
- ❏ Foundations with Break Lines
- ❏ Mechanical Space
- ❏ Existing Grade through Building
- ❏ HVAC
- ❏ Antenna
- ❏ Trim
- ❏ Canopies
- ❏ Awnings
- ❏ Earthquake Control
- ❏ Shafts
- ❏ Elevators
- ❏ Detail-Reference Indicators
- ❏ Wall-Section Reference
- ❏ _____
- ❏ _____
- ❏ _____

WALL SECTIONS

Review wall sections for compliance with codes, owner program, and the following:

- ❑ Demolition of Existing Conditions
- ❑ Masonry Coursing
- ❑ Building Systems and Assemblies
- ❑ Ceiling Cavity
- ❑ Materials and Finishes
- ❑ Insulation
- ❑ Post and Beam
- ❑ Panel
- ❑ Roof/Wall Conditions
- ❑ Canopy
- ❑ Awnings
- ❑ Site Drainage at Building
- ❑ Wall and Slab Waterproofing
- ❑ Architectural Shapes
- ❑ Projections
- ❑ Trim
- ❑ Casework
- ❑ Finished Grade at Building
- ❑ Light Wells below Grade
- ❑ Gratings
- ❑ Concealed Items in Wall Cavity
- ❑ Expansion and Contraction
- ❑ Detail References
- ❑ _____
- ❑ _____
- ❑ _____

- ❑ Vertical Dimensional Control
- ❑ Opening Dimensions
- ❑ Ceiling Assembly
- ❑ Skylights and Lightwells
- ❑ Fire Ratings
- ❑ Structure
- ❑ Bearing Wall
- ❑ Floor/Wall Construction
- ❑ Parapets
- ❑ Signs
- ❑ Foundation Conditions
- ❑ Below-Grade Waterproofing
- ❑ Waterstops
- ❑ Coves
- ❑ Ornamentation
- ❑ Millwork
- ❑ Specialties
- ❑ Existing Grade through Building
- ❑ Railings
- ❑ Ceiling Cavity
- ❑ Roof Penetrations
- ❑ Earthquake Control
- ❑ Consultant Coordination
- ❑ _____
- ❑ _____
- ❑ _____

BUILDING DETAILS

Review building details for compliance with codes, owner program, and the following:

❏ Demolition of Existing Conditions

❏ Cast-in-Place Concrete

❏ Metals

❏ Wood

❏ Plazas over Occupied Space

❏ Windows

❏ Curtain Walls

❏ Installation Conditions

❏ Earthquake Control

❏ Flashing

❏ Attachment/Anchoring

❏ Structural Analysis

❏ _____

❏ _____

❏ _____

❏ _____

❏ Dimensional Control

❏ Masonry

❏ Stucco

❏ Roofs

❏ Decks

❏ Storefronts

❏ Doors

❏ Expansion and Contraction

❏ Water Integrity

❏ Caulking and Sealants

❏ Hidden/Visible Attachment/Anchoring

❏ Consultant Coordination

❏ _____

❏ _____

❏ _____

❏ _____

DETAIL PLANS

Review detail plans for compliance with codes, owner program, and the following:

❑ Demolition of Existing Conditions

❑ Reference to Grid Systems

❑ Equipment Designations

❑ Tread Dimensions

❑ Railing Extensions

❑ Shaft Sizes

❑ Toilet Rooms

❑ Partitions

❑ Casework

❑ Wall Units Dotted

❑ Chase Dimensions

❑ Consultant Coordination

❑ Horizontal Dimension Control

❑ Interior-Elevation Reference

❑ Stairs

❑ Nosings

❑ Elevators

❑ Spreader Beams

❑ Fixtures

❑ Accessories

❑ Outline of Countertops

❑ Floor Drains

❑ Detail References

❑ _____

INTERIOR ELEVATIONS

Draw where needed to show material and color change and to show vertical relationships of items on the wall. Always draw casework/millwork elevations. Review interior elevations for compliance with codes, the owner program, and the following:

❑ Demolition of Existing Conditions

❑ Accessibility Standards

❑ Expansion and Contraction

❑ Display Devices

❑ Tackboards

❑ Casework Configurations

❑ Adjustable Shelving

❑ Toilet-Room Accessories

❑ Valances

❑ Consultant Coordination

❑ Vertical-Control Dimensions

❑ Design Issues

❑ Earthquake Control

❑ Chalkboards

❑ Markerboards

❑ Countertops

❑ Islands

❑ Soffets

❑ Group I/Group II Equipment

❑ _____

REFLECTED CEILING PLANS

Review reflected ceiling plans for compliance with codes, owner program, and the following:

❏ Demolition of Existing Conditions

❏ Grid References

❏ Walls Which Penetrate Ceiling

❏ Openings in Ceiling

❏ Lightwells

❏ Exposed Structure

❏ Ceiling Heights

❏ Drapery Tracks

❏ Material/Finish Indications

❏ Gypsum Board

❏ Metal

❏ Fixtures/Air-Handling Devices

❏ Surface-Mounted Items

❏ Earthquake Control

❏ _____

❏ _____

❏ _____

❏ Horizontal Dimensional Control

❏ Ceiling Steps

❏ Walls Which Abut Ceiling

❏ Skylights

❏ Stairs

❏ Access Panels

❏ Architectural Elements

❏ Sliding-Door Tracks

❏ Ceiling Grid

❏ Plaster

❏ Wood

❏ Recessed Items

❏ Expansion and Contraction

❏ Consultant Coordination

❏ _____

❏ _____

❏ _____

INTERIOR DETAILS

Review interior details for compliance with codes, owner program, and the following:

☐ Demolition of Existing Conditions

☐ Horizontal and Vertical References

☐ Doors

☐ Hardware

☐ Walls and Partitions

☐ Fire Ratings

☐ Materials/Finishes

☐ Casework

☐ Configurations

☐ Attachment/Anchoring

☐ Earthquake Control

☐ _____

☐ _____

☐ _____

☐ _____

☐ Dimensional Control

☐ Schedules

☐ Frames

☐ Relites

☐ Materials

☐ Acoustical Ratings

☐ Transitions

☐ Equipment

☐ Installation Conditions

☐ Expansion and Contraction

☐ Consultant Coordination

☐ _____

☐ _____

☐ _____

☐ _____

RECOMMENDED READING

Guidelines Newsletter. *Guidelines,* 18 Evergreen Drive, Box 456 Orinda, California 94563-0465
(415) 254-9393

Guzey, Onkal, and Freehof, James. *ConDoc.*
The New System For Formatting and Integrating Construction Documentation.
A Professional Development Program by The American Institute of Architects.

Construction Specification Institute. *Manual of Practice.* published by Construction Specification Institute.

Recommended Standards POP Manual. Northern California Chapter American Institute of Architects.

Guimmo, Vince. *Photoreproduction As A Drafting Tool.* Marketing Education Specialist. Drafting and Reproduction Technology. Eastman Kodak Company.

Guimmo, Vince. *Overlay Drafting Techniques.* Marketing Education Specialist. Drafting and Reproduction Technology. Eastman Kodak Company.

AUTHOR NAME. "Drawing From Logic". *Architecture* December, 1990. PAGE NUMBERS.

AIA DOCUMENT D200 PROJECT CHECKLIST. Published by The American Institute of Architects. 1982.

AIA DOCUMENT A111, STANDARD FORM OF AGREEMENT BETWEEN OWNER AND CONTRACTOR where the basis of payment is the COST OF THE WORK PLUS A FEE with or without a Guaranteed Maximum Price. Published by The American Institute of Architects, 1987.

AIA DOCUMENT B162, SCOPE OF DESIGNATED SERVICES. Published by The American Institute of Architects. 1977.

Vendor Catalogs
One of the best sources of information is vendor literature. Every manufacturer of drafting and reprographic supplies and equipment makes this information available, usually ordered by telephone. Some companies offer seminars or will come to your office to provide in-house training.

Index

About the Author

Thomas W. Berg is the Quality Control Manager for the architectural firm of Bassetti Norton Metler Rekevics in Seattle, Washington. He has over 25 years of experience in commercial, institutional, recreational, residential, and industrial architecture as both leading technical architect and project manager. He has worked with both public and private clients under conditions of design-build and fast-track construction. He is a past chairperson of the Winslow Planning Agency where he served two terms; was on the Winslow Traffic Advisory Committee; and served as curriculum advisor to the North Kitsap School Board of Education. He is on the Board of Directors of the Housing Resources Board in Bainbridge Island, Washington and The Board of Directors of the Association for Project Managers in the Design Professions (APM). His education is in engineering from St, Norbert College in DePere, Wisconsin, and he has been a licensed architect in the state of Washington since 1985.